The Laws of Scientific Change

Hakob Barseghyan

The Laws of Scientific Change

 Springer

Hakob Barseghyan
Institute for the History and Philosophy
    of Science and Technology
University of Toronto
Toronto, ON, Canada

ISBN 978-3-319-36786-6   ISBN 978-3-319-17596-6   (eBook)
DOI 10.1007/978-3-319-17596-6

Springer Cham Heidelberg New York Dordrecht London
© Springer International Publishing Switzerland 2015
Softcover reprint of the hardcover 1st edition 2015

Printed on acid-free paper

Springer International Publishing AG Switzerland is part of Springer Science+Business Media
(www.springer.com)

To my parents and grandparents

Acknowledgements

The project of this scope presupposes many intellectual debts. I express my sincere gratitude to

– *Levon-Harutyun Abrahamyan*, my great teacher, who introduced me to the world of philosophy; this project would not have been possible without his guidance.
– *Brian Baigrie*, my wonderful supervisor, who was open minded enough to think that a full-fledged theory of scientific change can be developed by a graduate student.
– *James Robert Brown* and *Paul Thompson*, who saw potential when it was far from obvious.
– *Craig Fraser, Nikolai Krementsov, William Seager, Mark Solovey, Marga Vicedo*, and *Denis Walsh*, my colleagues and friends, for their unconditional support and great advice.
– *Chris Doyle* and *Kristine Grigoryan*, my dear friends, who read the first draft of the manuscript and helped to improve the text immensely.
– *Joel Burkholder, Michael Fatigati, Robert Fraser, Rory Harder, Craig Knox, Gavin Lee, Parisa Moosavi, Andrew Oakes, Gregory Rupik, Clara Steinhagen*, and *Felix Walpole*, my students, friends, and fellow researchers, who scrutinize the theory and keep making it better.
– *Lucy Fleet*, my fantastic editor, who believed.

Contents

Introduction

It is science, more than any other aspect of modern culture that has advanced by insisting on the primacy of the general over the particular. The progress of modern science confirms the wisdom of those early investigators who dared to ignore immediate and obvious differences while reaching for general truths.

Larry Laudan

At any moment of the history of science, there are certain *theories* that the scientific community of the time accepts as the best available descriptions of their respective domains. In addition to accepted theories, there are also certain *methods* employed by the scientific community in theory assessment. This set of all accepted theories and all employed methods constitute what can be called the *scientific mosaic* of the time. Now, it is obvious that the scientific mosaic is in a process of constant change. Theories that were accepted in the past may or may not be part of the contemporary scientific mosaic. For instance, nowadays, the scientific community accepts that the best available description of physical processes is provided by general relativity (with the Big Bang cosmology based on it) and the Standard Model of quantum physics. Yet, a mere century ago, it was accepted that the best description of physical processes was provided by such theories as Newtonian mechanics and Maxwellian electrodynamics, while some 400 years ago, it was commonly accepted that the world was best described in terms of the Aristotelian-medieval natural philosophy. The same holds for the employed methods: most of the contemporary criteria actually employed in theory assessment have little in common with the requirements employed in the Aristotelian-medieval period. Thus, it is safe to say that the process of scientific change concerns not only the *theories* accepted by the community but also the *methods* that the community employs in theory assessment.

The central question of this book is whether there are any *laws* governing the process of scientific change. Is the process of transitions from one accepted theory to the next and from one employed method to the next completely random, or does it obey certain general laws? Are there any general patterns in this process? In other words, can there be a *general descriptive* theory of scientific change?

Contrary to the widespread opinion, I argue and that there can be a *general descriptive theory of scientific change* (TSC). Thanks to a growing body of historical research, we are currently in a position to say that scientific change *is* a

law-governed process. Despite all the apparent dissimilarities of different historical episodes, there are universal patterns of theory acceptance and method
employment. While it is possible for two mosaics to have virtually nothing in
common, *changes* in these mosaics are always governed by the same set of laws.
Thus, I oppose the *particularist* approach of the disunity of science movement,
which holds that the process of scientific change cannot possibly be explained by
any general theory. I believe that a general descriptive theory of scientific change
can and *does* exist.

According to the *particularist* approach, the process of scientific change can
only be studied in a fragmentary fashion and, strictly speaking, no general theory
about the mechanism of scientific change is possible. It has been argued that one
should avoid any general theories since there is always a risk of "shoehorning history" into the schemes of this or that general theory.[1] The whole idea of finding
regularities in history has been labeled "whiggish" and is believed to be inevitably
distorting the actual historical episodes.[2] There are several lines of reasoning which
allegedly back up this position. For one, there have been several unsuccessful
attempts of developing a general TSC; recall, for instance, theories of Popper, Kuhn,
Lakatos, or Laudan, which clearly failed to describe the actual process of scientific
change. Moreover, the complexity and apparent disunity of historical episodes
revealed by recent studies do not seem to be readily explainable by existing theories
of scientific change. Add to this the contemporary wisdom that historical episodes
are so immersed in their local sociocultural contexts that no general theory whatsoever can possibly account for all of them. As a result, the particularist position
appears virtually unavoidable.

But is particularism inevitable, or is there after all a way to construct a *non-
whiggish* TSC? Can there be a general TSC that does justice to historical episodes
and does not shoehorn them into faulty templates? It is my belief that such a general
TSC is possible, for the position of particularism is ill founded.[3] Particularism is
based on a simple but faulty assumption once taken for granted by virtually all philosophers of science. The tacit assumption is that explaining the mechanism of scientific change amounts to explicating the universal and unchanging method of
science, which presumably guides transitions from one accepted theory to the next.
On this view, any general TSC presupposes the existence of the unchangeable and
universal method of theory appraisal. This tacit assumption was implicit in the theories of logical positivists, Popper, Lakatos, and the early Laudan, among many others. Thus, it is not surprising that the project of general TSC suffered a serious blow
once it became clear that there is no such thing as the unchangeable and universal
scientific method. It was this recognition of the disunity and changeability of methods that led to the current state of affairs, where the project of creating a general

---

[1] Allchin (2003), p. 315. For discussion, see Nickles (1995).

[2] For the idea of "whig history", see Butterfield (1931); Stockings (1965); Hull (1979); Wilson and
Ashplant (1988a, b); Mayr (1990).

[3] Luckily, some contemporary authors also seem to agree. As Buchwald and Franklin have put it,
"the pendulum has swung too far". Buchwald and Franklin (2005), p. 1.

TSC is virtually abandoned. And that is where we currently stand: no universal method – no general theory of scientific change.

Yet, I will argue that this conclusion is false, for it is based on a false premise. A general TSC, as defined in this book, is a descriptive theory that aims at explaining changes in both accepted *theories* and employed *methods*. It does not presuppose the existence of any fixed and universal method of science and thus does not attempt to portray all cases of theory change as being guided by some fixed and universal method. It is true that scientific change concerns not only theories but also methods, for different time periods and different fields of inquiry often employ quite different criteria of theory appraisal. However, the changeability of method does not necessarily imply the impossibility of a general TSC. What it does imply is that any adequate TSC should account not only for changes in accepted *theories* but also for changes in employed *methods*.

There is good reason to believe that a general TSC is not only theoretically possible but also practically achievable. Indeed, no serious scientist would ever take initial failures in finding regularities as a reason for despair. Physicists, for instance, do not stop searching for general laws when their initial attempts fail to produce the desired results. Likewise, no level of complexity or apparent disunity of historical episodes can justify the particularist abolition of the idea of a general TSC. In fact, no matter how dissimilar a falling stone and a moving planet may appear in observations, only a physical *theory* can tell us whether their behaviour is governed by different laws (as Aristotle would have it) or the same laws (as Descartes and Newton would have it). Similarly, in order to tell in what respects two historical episodes are similar or dissimilar, one needs a general TSC. It is of course true that the outcome of a historical episode depends crucially on its specific historical context. Yet, it is quite likely that different historical contexts are simply different initial conditions that enter into the same set of equations. After all, it does not surprise us when the same physical theory accounts for a vast number of apparently dissimilar phenomena in seemingly different 'physical contexts'. Do a magnet and a light ray look similar? Not at all! And yet, we do believe that their behaviour is described by the same set of laws and that the difference between them is in the respective *initial conditions*, not the *laws* themselves. Likewise, there is nothing whiggish in interpreting differences in historical contexts as differences in initial conditions and not the governing laws. This is what a proper non-whiggish general TSC will attempt to do.

A general theory of scientific change is not only possible but also highly desirable. After all, if there is any lesson that we have learnt studying the history of science, it is that there are no pure statements of fact, for even the most basic observational propositions presuppose some general assumptions. This has come to be known as the thesis of *theory-ladenness* of observations. The history of any scientific discipline is full of illustrations of theory-ladenness. Take a straightforward example – a falling stone. Where Aristotle saw a heavy body descending towards the centre of the universe, Descartes and Huygens saw a displacement of a slower object by swifter particles in the terrestrial vortex. Where Newton saw a gravitational acceleration, Einstein saw an inertial motion in a curved spacetime.

The history of science teaches that no phenomenon can receive a theory-free interpretation, for any interpretation presupposes, if not a full-fledged theory, then at least individual general propositions. In short, it is obvious that no science can proceed without general propositions, and the history of science is no exception. It is true that 'the historian seldom makes explicit use of historical generalizations'.[4] In our historical narratives, we often leave these general assumptions tacit. Yet, they are there all the same. Thus, we face a dilemma: we can either continue relying on *tacit* general assumptions about the process of scientific change or we can opt to formulate them *explicitly* even if they seem trivial at first and begin to subject them to scientific scrutiny. If we choose to formulate our assumptions openly, we will be able to criticize them, trace and evaluate the logical connections among them, and eventually bring them into a coherent system of propositions of a general TSC. This TSC will then provide us with necessary tools to explain different historical episodes in a systematic fashion by means of a unified vocabulary, just as the laws of theoretical physics help to make sense of physical experiments and observations or just as the laws of biology help to reconstruct the tree of life. I believe it is time for us to build on the success of the history of science and proceed to the next step – formulation of a general descriptive theory that would uncover the mechanism of scientific change. Several decades of fascinating historical scholarship have prepared a great starting point for this crucial step.

The problem is, however, that at the moment, there is no consensus as to what a general TSC should be, what kind of phenomena it has to explain, and what criteria it must satisfy. Part of the ambiguity stems from the vagueness of our basic historical vocabulary. For instance, when talking about the stances that the scientific community can take towards a theory, we often use such phrases as 'accept', 'universally receive', 'pursue', 'embrace', or 'acknowledge'. Are all these terms synonyms, or are there subtle differences among them? Does 'the theory was accepted circa 1800' mean the same as 'the theory was pursued circa 1800'? And if there are important differences, then changes in which of these stances should a TSC trace and explain? Obviously, it is impossible to answer this last question unless we first spend some time clarifying our basic terminology. Among other commonly confused terms are 'method' and 'methodology', 'justification' and 'appraisal', 'construction' and 'discovery'.

This all contributes to the ambiguity regarding the *scope* of a TSC. Is it *normative*, is it *descriptive*, or is it *both*? Should it account for changes in *individual* belief systems, or should it account for changes in theories accepted by the *scientific community*? Should it explain changes in *all* fields of science, or only those in *natural science*, or only those in *physics*, etc.? What *time period* should it cover – only the last 100 years, the period since the seventeenth century, or all changes since the time of Aristotle? Should it explain how theories become *accepted*, or should it explain how theories are *constructed*, or should it explain both? It is safe

---

[4] Bunge (1998), p. 290.

to say that we are yet to provide satisfactory answers to these questions. Not only do we lack any workable TSC, we don't even realize what we *expect* from such a theory. Thus, before we can proceed to constructing an actual TSC, we must clarify its scope; the question of the scope of a general TSC is tackled in the first chapter of this book.

Once we have clarified the scope of the project, we must proceed to the next question: how is such a theory *possible*? It is obvious that there is no point in debating about the possibility or impossibility of a general TSC unless we first clarify what this theory purports to be. It is ironic that those who ardently deny the very possibility of such a theory virtually never bother to specify *what* exactly they deny; they hardly ever clarify what they mean by a general TSC. Only when we understand what the scope of the project is can we discuss the arguments for and against its feasibility. The main arguments against the possibility of a general TSC are discussed in the second chapter of this book.

Finally, when we have both clarified the scope of the project and ascertained its possibility, we must address the question of its *assessment*. Suppose we have succeeded in constructing a TSC. How are we going to evaluate (assess, test) it? What conditions should a TSC satisfy in order to become accepted? We must answer this question before we can legitimately move on to constructing an actual TSC. The question is discussed in the third chapter, called *Assessment*.

These three chapters together compose what I call the *metatheory* of scientific change, i.e. a theory that addresses the issues of the scope, possibility, and assessment of a theory of scientific change. Thus, *Part I* of the book has the following structure:

Part I. Metatheory		
1. Scope	**2. Possibility**	**3. Assessment**
? What is the *scope* of a general theory of scientific change?	? How is a general theory of scientific change *possible*?	? How is a general theory of scientific change to be *evaluated*?

It must be appreciated that these preliminary metatheoretical discussions are indispensable. When the physicist sets off to construct her physical theory, normally she can skip any discussions about the scope, possibility, or assessment of physical theories in general. She can directly proceed to building her theory, for there are already certain accepted views on what physical theories should account for, how they are possible, and how they must be evaluated. In other words, the physicist is in a position to start with a theory proper, for there is some tacitly accepted metatheory of physics. We, on the other hand, cannot start with a theory proper, for we lack an accepted metatheory of scientific change. Only when we have clarified what exactly a general TSC should explain, how it is possible, and how it is to be assessed can we proceed to constructing an actual theory of scientific change.

The actual *theory* of scientific change is developed in *Part II* of the book. It is presented in an axiomatic form, where the axioms are postulated and the theorems are deduced from the axioms. Thus, the second part includes two chapters:

Part II. Theory	
4. Axioms	**5. Theorems**
? What are the *axioms* of the general theory of scientific change?	**?** What *theorems* can be deduced from the axioms of the theory?

Albeit unconventional, this form of presentation is the most effective, especially when applied to complex theories. It is easier to deal with a complex system of propositions when their logical relations are explicitly stated. I, for one, would most certainly be unable to keep track of the whole system if it were not put in an axiomatic form. In addition, explicit formulation of all axioms and theorems has proven conducive to constructive criticism and further elaboration of a theory. It helps to adopt a piecemeal approach when each individual deduction is scrutinized separately. This strategy helps to ensure that the task of the theory's further improvement is bearable.

One final remark before we set off: The practice shows that when dealing with a complex topic such as this, it is helpful to employ visual diagrams in order to keep things simple. Thus, the text is full of diagrams. Although the first few diagrams may seem redundant, eventually they will prove useful. While in the beginning, diagrams simply restate what is clearly explained in the text, they gradually become more and more informative. Finally, in the later chapters, they become indispensable – they play a crucial role of complementing the text and ensuring that the text is understood correctly. To this end, all the diagrams are constructed by means of a special system of symbols. Where possible, I employ standard symbols from the *Unified Modeling Language* (UML). Since UML does not contain any special symbols for logical relationships (such as conjunction, disjunction, implication, etc.), I have taken the liberty of introducing some additional symbols. However, no prior knowledge of UML is required as the meaning of each symbol will be explained upon first use. For the reader's convenience, there is also a legend at the end of the book.

Part I
Metatheory

Chapter 1
Scope

The first task at hand is to clarify the scope of the project by answering the following questions. Should a theory of scientific change be *normative* or *descriptive*? Should it deal with the process of theory *construction* or the process of theory *assessment*? Should it trace changes from one *accepted* theory to the next, from one *used* theory to the next, or from one *pursued* theory to the next? Should it explain changes in belief systems of *individual* scientists or should it focus on the scientific *community*? Should it explain transition in *implicit* rules of theory assessment actually employed by scientists or should it explain changes in *explicitly* formulated rules? Finally, changes at what *time periods* and in what *fields* of science should it explain?

Theories and Methods

It is safe to say that, at any moment of time, there are certain theories that the scientific community of the time takes as the best available descriptions of their respective domains. In my definition, *"theory"* may refer to any set of propositions that attempt to describe something. Theories may be empirical (e.g. theories in natural or social science) or formal (e.g. logic, mathematics). Theories may be of different levels of complexity and elaboration, for they may consist of hundreds of systematically linked propositions, or of a few loosely connected propositions.[1] They may or may not be axiomatized, formalized, or mathematized. Our current scientific mosaic, for instance, includes such highly developed and complex theories as general relativity, the Standard Model, and the theory of biological evolution. But it also includes many individual general propositions, such as *the law of diminishing returns* in economics or the epistemic conception of *fallibilism*, which are only loosely linked

---

[1] We may therefore use "theory" and "proposition" interchangeably, since any proposition is essentially a folded theory, while any theory is a set of propositions or, in the extreme case, a single proposition.

© Springer International Publishing Switzerland 2015
H. Barseghyan, *The Laws of Scientific Change*, DOI 10.1007/978-3-319-17596-6_1

to other accepted propositions.[2] Finally, theories may be composed of both *general* and *singular* propositions. Indeed the contemporary scientific mosaic includes not only general propositions of physics, chemistry, or biology, but also singular propositions of history. The important point is that, at any moment of time, there is a set of theories that the scientific community takes as the best available descriptions of their objects.

It has often been argued that theories are best construed not as propositions but as *models* which are abstract set-theoretic entities. Importantly, on this *model-theoretic* or *semantic* view of theories, models do not contain propositions but are structures of non-linguistic elements.[3] Whether this is indeed the case is to be established not by this metatheory but by an actual theory of scientific change. What is important from our perspective is that even on this model-theoretic view acceptance and rejection of theories depends crucially on formulating descriptive propositions.[4] For something to become accepted as true or truthlike it must be expressible in propositions at least *in principle*. Take an example of the Aristotelian-medieval model of the cosmos. When the medieval scientific community accepted this model, the community essentially accepted a tightly connected set of *propositions*, such as "the Earth is in the centre of the universe", "the Moon, the Sun and all other planets are embedded in concentric crystalline spheres which revolve around the central Earth", "all celestial bodies are made of element aether", "aether is indestructible", "aether has a natural tendency to revolve around the centre of the universe", "all terrestrial bodies are made of the four terrestrial elements", etc. In short, while models may as well play an important role in scientific practice, no part of these models can be actually accepted or rejected if it is not expressible in propositions. The same holds for propositions that are not openly formulated but are accepted tacitly. What matters is that *in principle* they too can be openly expressed in propositions. In general, if something is not expressible as a proposition, then it cannot have a truth value and cannot be accepted or unaccepted as the best description of anything. Thus, from the perspective of our project, it is safe to treat theories as collections of propositions.

In addition to accepted theories, there are also methods which the scientific community of the time employs in theory assessment. It is worth mentioning that traditionally the word "*method*" has had two major connotations. In one sense, it has referred to criteria (rules, standards) of theory *assessment* (evaluation, appraisal). In the other sense, it has referred to techniques for investigation or heuristic guidelines for theory *construction* (generation, invention). Unfortunately, there's been a long-standing tradition of confusing the two connotations.[5] In order to avoid confusion, I shall use "method" only in the former sense: in my definition, "method" is a set of

---

[2] This issue is addressed in detail in section "Time, Fields, and Scale" below.

[3] See Suppe [1989].

[4] For a thorough discussion and criticism of the model-theoretic view of theories, see Chakravartty (2007), pp. 188–205, especially pp. 192–199.

[5] Hacking, for instance, uses "method" to denote *heuristic procedures for theory construction* or *research procedures*, while his term for *criteria of theory appraisal* is "logic". See his (1996), pp. 51, 54, 64, 66.

criteria for employment in theory assessment.[6] It is obvious that different methods may have different applicability. While some methods may be of general nature and apply to all scientific disciplines, other methods may be very specific and apply only to a particular field of inquiry. For instance, a requirement that "a theory is acceptable only if it has confirmed novel predictions" can, in principle, be applicable to all scientific disciplines.[7] In contrast, a requirement that "a hypothesis about a drug's efficacy is acceptable if the drug's effect has been confirmed in a double-blind trial" is applicable only to drug efficacy hypotheses.

In short, at any moment of time, there are certain accepted theories and certain methods employed in theory assessment. This set of all accepted theories and employed methods constitute what I shall call the *scientific mosaic* of the time. Here are our basic definitions[8]:

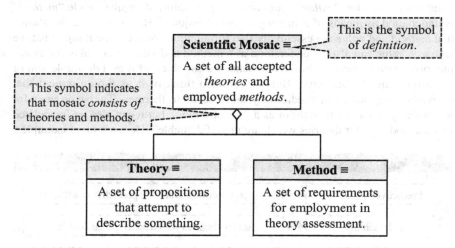

The reason why I call this set of theories and methods "mosaic" and not, say, "system" is that the elements of the mosaic may or may not be tightly adjusted; there may be considerable gaps between the elements of the mosaic. For instance, nowadays we realize that there is a considerable gap between general relativity and quantum mechanics and, yet, we do not hesitate to accept both.

Obviously the scientific mosaic is in a process of perpetual change. Most of the theories that we accept nowadays didn't even exist 200 or 300 hundred years ago. Similarly, at least some of the methods that we employ in theory assessment

---

[6] For detailed discussion, see section "Construction and Appraisal" below.

[7] Whether this requirement is or has been *actually* employed in theory assessment is a specific historical issue. For my attempt to explicate one aspect of the implicit requirements currently employed in theory assessment, see *Part II*, section "The Third Law: Method Employment", pp. 142 ff.

[8] It is of course conceivable that the mosaic may also contain *normative* propositions, such as those of ethics (e.g. "racial discrimination is unacceptable"). Yet, the status of normative propositions is currently debatable and that is the reason why, at this point, the concept of *theory* only includes *descriptive* propositions.

nowadays have nothing to do with the methods employed in the seventeenth century. This brings us to our definition of *scientific change*:

> **Scientific Change ≡**
>
> Any change in the *scientific mosaic*, i.e. a
> transition from one accepted *theory* to another
> or from one employed *method* to another.

Some authors have suggested that the process of scientific change involves not two but three different classes of elements. Thus, according to Larry Laudan, scientific change involves not only theories and methods, but also *epistemic values*. In Laudan's conception, "*value*" is defined as a goal (aim) of inquiry, while "*method*" is understood as means of achieving a goal of inquiry.[9] However, it can be shown that *method* and *value* are essentially two formulations of the same thing: what we take as a value can also be understood as a method and vice versa. Take, for example, *predictive accuracy*. On the one hand, we can think of it as a desirable aim of scientific inquiry, i.e. value. But we can also think of it as a requirement that prescribes acceptance of predictively more accurate theories. The same goes for *falsifiability*. It can be thought of as a goal of inquiry, but it can also be formulated as a method: "prefer theories which are more falsifiable". Here are more examples:

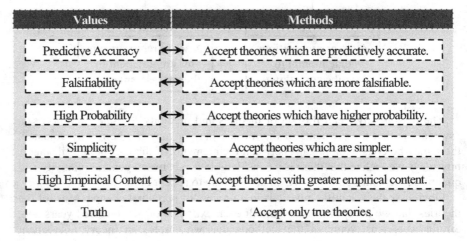

Values	Methods
Predictive Accuracy	Accept theories which are predictively accurate.
Falsifiability	Accept theories which are more falsifiable.
High Probability	Accept theories which have higher probability.
Simplicity	Accept theories which are simpler.
High Empirical Content	Accept theories with greater empirical content.
Truth	Accept only true theories.

As we can see, this applies even to such abstract values/methods as *truth*. To say that truth is the goal of scientific inquiry is the same as to say that only true theories must be accepted. Similarly, saying that the goal of science is greater approximation to truth amounts to saying that a theory is acceptable only if it is the best available approximation to truth. Of course, "accept only the best available approximations to truth" method is extremely abstract, for it doesn't address the key question of what

---

[9] See Laudan (1984), Laudan (1996), pp. 132–133.

makes a theory the best available approximation to truth. However, strictly speaking it *is* a method and, thus, must be treated as such.

I believe the same holds for any value/method pair. The best indication of this fact is that what one author regards as a value another takes as a methodological rule and vice versa. For instance, while Kuhn treats *simplicity* as a value, Lycan treats it as a method.[10] Similarly, Laudan himself never explains why *predictive accuracy* is taken as a goal, while *internal consistency* is taken as a method.[11] Why are they arranged the way they are and not vice versa? When discussing the history of methodological controversies in the eighteenth century, Laudan presents the debate between inductivism and hypothetico-deductivism as an instance of axiological debate regarding values. He provides no explanation as to why this is an example of axiological and not methodological debate.[12]

In short, there is no rationale underlying the method/value distinction. Yes, the two can be defined differently; it is all in our hands. Yet, when it comes to actual methods/values, it becomes clear that what can be formulated as a value (goal), can be also formulated as a method (requirement, criteria). That is why we don't need three classes of elements where we can easily manage with two – theories and methods of their assessment; we have certain beliefs about the world and its workings and we also have certain expectations as to how these beliefs are to be assessed.

The main reason why Kuhn and Laudan think values and methods belong to two different classes is that often more *concrete* methods are means to satisfying the requirements of more *abstract* methods. In other words, methods come in different degrees of concreteness/abstraction. Nowadays, for example, we require that new drugs should be tested in double-blind trials, so the method stipulates that "a hypothesis concerning a drug's efficacy is acceptable if it is confirmed in *double-blind* trials". By meeting the requirements of this concrete method, we simultaneously satisfy a bit less concrete requirement that "a hypothesis concerning a drug's efficacy is acceptable if it is confirmed in a controlled trial", since double-blind trial is species of controlled trial. This second requirement itself is a means of meeting a somewhat more abstract requirement that "a hypothesis concerning the existence of new causal relations is acceptable if it is confirmed in *repeatable* experiments and observations". This latter requirement is a specification of a more abstract requirement that "a new hypothesis that suggests the existence of new causal relations is acceptable if it is *confirmed* in experiments and observations" (this one doesn't include the repeatability clause). This goes all the way up to the most abstract requirement that "a hypothesis is acceptable if it is the best available description of its object". As a result, at any moment of time, there is a *hierarchy* of methods, where the requirements of more concrete methods specify the requirements of more

---

[10] See Kuhn (1977), p. 331 and Lycan (1988), p. 130. Nola and Sankey have pointed this out in their (2000), p. 11.

[11] See Laudan (1984), pp. 31, 91. I owe this example to Knowles (2002), p. 173.

[12] See Laudan (1984), pp. 55–61. Knowles has also pointed this out in his (2002, p. 184, footnote 4).

abstract methods; by satisfying more concrete requirements, a theory also satisfies some more abstract requirements.[13] Thus:

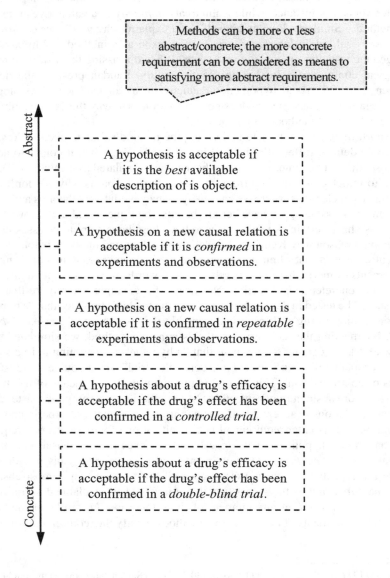

---

[13] The existence of this hierarchy has been emphasized by Robert Fraser during the seminar of 2014. The idea that criteria of theory evaluation can be of different level of abstraction/concreteness is also implicit in Hansson's account of demarcation. See Hansson (2013).

I think it is this hierarchic structure of methods that has confused both Kuhn and Laudan: since some requirements are means to satisfying some other (more abstract) requirements, Kuhn and Laudan mistakenly thought that there is a two-level hierarchy where methods are means to achieving certain ends (values). Yet, in reality, there can be more than two levels in this hierarchy and, importantly, each element of this hierarchy is a method, i.e. a requirement to be employed in theory assessment. There is simply no need in a separate class of values; Kuhnian and Laudanian values are nothing but more abstract methods. Suffice it to acknowledge that at any moment of time the employed methods constitute a hierarchy of requirements.[14]

Thus, it is safe to say that the process of scientific change involves theories and methods. Changes in the scientific mosaic can be viewed as a series of successive frames, where each frame represents a state of that mosaic at a given point of time. Obviously, such a frame would include all accepted theories and all employed methods of the time. Schematically, it may be portrayed thus (note that new elements of the mosaic are shaded):

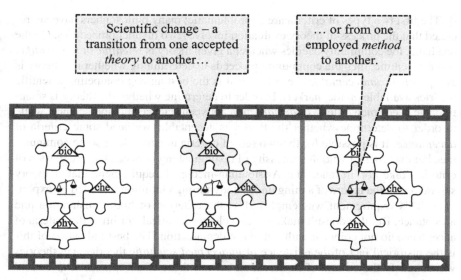

Now, the definition of method suggests that it can consist of different criteria for theory assessment. It is readily seen that these criteria come in different types and perform different functions. The purpose of some criteria is to determine whether a given theory is *scientific* or *unscientific*. Other criteria are supposed to tell us whether a given theory is *acceptable* or *unacceptable*. Yet other criteria help to establish the mutual *compatibility* or *incompatibility* of two given theories. Thus, a method can consist of the requirements of at least three different types[15]:

---

[14] Needless to say that this hierarchy can be very different at different time periods.

[15] I do not claim that this list is exhaustive; it is quite possible that there are other classes of criteria which perform other functions. That is for an actual TSC to establish in collaboration with HSC.

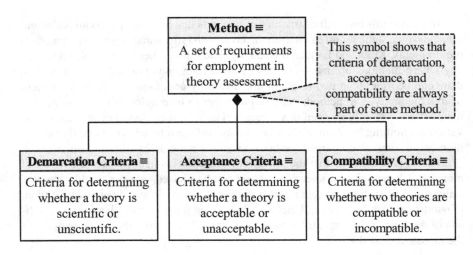

The first two types of criteria are unproblematic; many philosophers have appreciated that the process of theory evaluation involves two distinct procedures. On the one hand, the community decides whether a given theory is *scientific* or *unscientific*. If it is scientific, then the community proceeds to determining whether the theory is *acceptable* or *unacceptable* (i.e. whether it is the best among competing scientific theories available on the market). In order to determine whether the theory is scientific, we need some criteria of *demarcation* between science and non-science, while in order to determine whether the theory is acceptable, we need some criteria of *acceptance*. It is possible for the two sets of criteria to coincide in some communities, but that doesn't void the necessity of distinguishing between the two classes of criteria. Take for instance the Aristotelian-medieval requirement that a theory should reveal the nature of a thing under study through intuition schooled by experience. This requirement was employed as the criterion of both demarcation and acceptance. But this doesn't make the two classes identical, for often the criteria of acceptance do not coincide with criteria of demarcation. The best indication of this is the historical fact of the existence of *unaccepted scientific* theories, i.e. theories which satisfy the criteria of demarcation but not those of acceptance. Our current attitude towards superstring theories is a good example: we believe that these theories are scientific, yet we do not believe that they are the best available descriptions of their domain. In short, from a logical standpoint, the criteria of demarcation and criteria of acceptance belong to two different classes.[16]

---

[16] This distinction has been appreciated by philosophers as diverse as Carnap and Popper. Of course, there is little agreement as to what these respective criteria stipulate and how they are to be spelled out, but it is commonly assumed that criteria of demarcation are not necessarily the same as criteria of acceptance: to say that a theory is scientific is one thing, to say that it is better than all of its scientific competitors is another thing. For outlines of respective positions see Uebel (2011) and Thornton (2009).

As for the criteria of compatibility, although their existence has never been seriously questioned, their function has been traditionally oversimplified. It has been often tacitly assumed that compatibility or incompatibility of any two theories is decided on purely logical grounds: if the two theories logically contradict each other, they are incompatible. Yet, there is reason to suspect that the actual situation is much more complex than this simplistic picture allows. The fact that we often simultaneously accept theories which strictly speaking logically contradict each other is a good indication that the actual criteria of compatibility employed by the scientific community might be quite different from the classical logical law of noncontradiction. Perhaps the most famous illustration of this phenomenon is the case of general relativity and quantum physics. While we accept both theories as the best available descriptions of their respective domains, we also know that strictly speaking the two theories contradict each other. For example, the contradiction becomes apparent when the two theories are applied to black holes. This seems to suggest that the criteria of compatibility employed by the scientific community do not coincide with the logical law of noncontradiction: we have found a way to combine the two theories in our mosaic despite the fact that they logically contradict each other. In addition, it is possible that at different time periods different communities have employed different criteria of compatibility. We can easily conceive two different communities, one of which sticks to the strict classical notion of consistency and requires that each element of their mosaic be always consistent with all other elements, while the other community employs a more flexible notion of compatibility which allows for two mutually inconsistent theories to be accepted in the same mosaic. Whether this is indeed the case can only be established by studying actual historical episodes. What we have to appreciate at this stage is that (1) the criteria of compatibility may turn out to be changeable and (2) they need not necessarily coincide with the logical law of noncontradiction. In brief, criteria of compatibility are part of the employed method and can turn out to be changeable just like any other criteria.

This brings us to the main question of this section: which classes of elements should a TSC be concerned with? The short answer is that it has to be concerned with changes in both theories and methods. This includes all the theories in the mosaic and all the methods regardless of their place in the hierarchy of methods or their function (demarcation, acceptance, compatibility). This follows directly from the definition of scientific change. This answer is very general and needs to be specified. That is what I shall do in the following sections of this chapter.

Descriptive and Normative

There are at least three different sets of questions concerning the process of scientific change – historical (empirical), theoretical, and methodological:

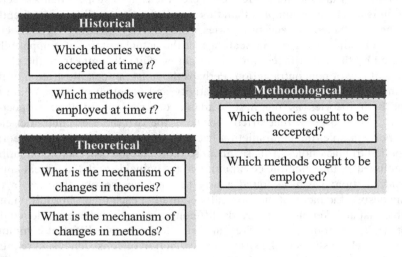

It is obvious that the historical and theoretical questions are essentially *descriptive*, while the methodological questions are *normative*.[17] Indeed, when the historian tries to reconstruct the state of the mosaic at a given time, she basically provides a description of that state. Similarly, when the theoretician of science attempts to understand the mechanism of scientific change, she is engaged in a purely descriptive enterprise; the theoretician attempts to explain how and why changes in the mosaic *actually* take place, without giving any prescriptions as to how those changes *ought to* take place. The latter is the task of the methodologist, who attempts to determine which methods ought to be employed in theory evaluation and which theories ought to be accepted.

Therefore we deal with three disciplines here – history of scientific change (HSC), theory of scientific change (TSC), and normative methodology (MTD). It is the task of MTD to evaluate the existing methods and tell us which of these methods ought to be employed in theory appraisal in the future. Reconstructing the state of the mosaic at different time periods is one of the main tasks of the discipline called HSC.[18] Finally, understanding the mechanism of changes in the mosaic of accepted

---

[17] In general, *normative* propositions say how something *ought* to be, what's good or bad, what's right or wrong. In contrast, *descriptive* propositions aim to describe or explain how things actually *are*, or how they were in the past, or predict how they are going to be in the future without any value judgement. It is important to keep in mind than, in this context, the category of *descriptive* includes both propositions that describe how things are and propositions that explain why things are the way they are.

[18] Traditionally, "history" has been used to denote both the actual process in time and the historical study of that process. To avoid confusion, henceforth, "the history of scientific change" (or HSC,

theories and employed methods is, as I shall argue, the task of TSC. In short, while HSC and TSC are descriptive disciplines, MTD is essentially normative:

Descriptive	Normative
History of Scientific Change ≡ A descriptive discipline that attempts to trace and explain *individual* changes in the scientific mosaic. **Theory of Scientific Change** ≡ A descriptive discipline that attempts to uncover the actual *general* mechanism of scientific change.	**Methodology** ≡ A normative discipline that formulates the rules which *ought to* be employed in theory assessment.

Although this distinction seems trivial, it is an unfortunate historical fact that nowadays it is not fully appreciated. In particular, there is currently no descriptive theory of scientific change to speak of. Instead, the descriptive issues concerning the general mechanism of scientific change and normative issues concerning the choice of best theories and methods are often conflated into one inextricable mixture. This conflation stems from a long-standing tradition which has assumed that to explain the mechanism of scientific change (descriptive) amounts to prescribing the best course of action (normative). The discipline that supposedly tackles this mixture of normative and descriptive issues has been labeled as "theory of scientific method", "theory of rationality", and "theory of scientific knowledge", or even "philosophy of science" to name a few of the labels. To put it mildly, these labels don't help to clarify the confusion but make matters only worse. Take "theory of scientific knowledge", for instance. How is this "theory" to be understood? Is it meant as a description of what *is* actually going on in science, or is it a prescription of what *ought to* be going on? The same applies to the most widespread of these labels – the notorious "philosophy of science". Traditionally, it has been understood as a mixed normative-descriptive discipline where "philosophers of science look at a given science and ask what is really going on here, and what is the best way to know it – and, in some instances are in a position to make recommendations about how best to conduct science in future."[19] The philosopher of science, on this account, is viewed both as a theoretical historian (descriptive) and as a methodologist (normative).

This unfortunate conflation can be traced back to William Whewell's *The Philosophy of Inductive Sciences*. According to Whewell's definition, "The Philosophy of Science … [is an] insight into the essence and conditions of all real knowledge, and an exposition of the best methods for the discovery of new truths."[20]

---

for short) will denote exclusively the study of the process of scientific change, while "the process of scientific change" will denote the object of that study.

[19] Pinnick and Gale (2000), p. 111.

[20] Whewell (1840), p. 1.

Thus, for Whewell, the philosophy of science is both descriptive and normative, for it is both a description of "the essence and conditions of all real knowledge" and "an exposition of best methods". Thomas Kuhn is apparently another source of the conflation. It is not at all clear whether his conception of paradigms and scientific revolutions is to be understood as a *description* of how scientists have *actually* proceeded in theory appraisal, or whether it is a *prescription* of how scientists *ought to* proceed.[21] Personally, Kuhn claimed that it "should be read in both ways at once".[22] Similarly, both Lakatos's methodology of scientific research programmes and the early Laudan's problem-oriented methodology were constructed both as *descriptions* of the actual workings of science and as *prescriptions* of what scientists ought to do.[23] Many contemporary authors working in the field inherit this view from Kuhn, Lakatos, Laudan and other classics of the genre.[24] Take, for instance, Robert Nola and Howard Sankey who say that the theory of scientific method is "about what allegedly *does*, or what *ought to* happen when one theory (paradigm, research programme or whatever) is followed by another."[25]

There is a straightforward line of reasoning that yields this conflation. It is, I believe, based on the belief that there is one unchangeable method of theory appraisal, *the* so-called scientific method:

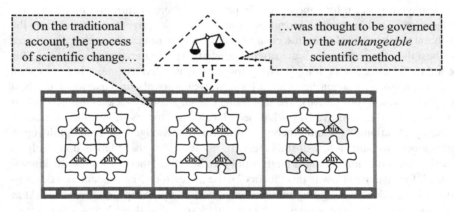

---

[21] See Nola and Sankey (2000), p. 28 (my *emphasis*).

[22] Kuhn (1970a), p. 237.

[23] The transition from the early Laudan to the later Laudan took place towards the late 1970s, when Laudan recognized that there are no unchangeable methods. While the main work of the early Laudan is his (1977), the main work of the later Laudan is his (1984). As far as I know, McMullin was the first to recognize this transition. See McMullin (1988), p. 15.

[24] It is known that Lakatos changed his mind towards the end of his life. In one of his final letters to Feyerabend, he admitted that he had converted to the view that standards (i.e. methods) of science are changeable. He also announced that he was going to write a book titled *The Changing Logic of Scientific Discovery*. See Motterlini (1999), pp. 355, 357. Unfortunately, Lakatos died of a heart attack less than a month after writing that letter and the book was never completed. Only one piece called "Newton's Effect on Scientific Standards" was written and was later published as chapter 5 of Lakatos (1978a). Thus, Lakatos authored only one full-fledged theory, presented in Lakatos (1970). For discussion, see Motterlini (2002).

[25] Nola and Sankey (2000), p. 8.

On this traditional account, the method of science was taken as something external to the process of scientific change, something unchangeable that guides the process of transitions from one theory to the next. Thus, the process of scientific change was thought to concern only theories, while the universal and fixed method of science was thought to be ahistorical. By accepting this view, the classics ended up conflating the descriptive theory of scientific change (TSC) with normative methodology (MTD).

Indeed, if we were to assume that the scientific method employed in theory appraisal is unchangeable (fixed, ahistorical), then we would be in no position to prescribe any other method for future use, since this same method would be the one that would have to be employed in all future theory assessments. In that case, descriptive TSC would become indistinguishable from normative MTD, for by answering the *descriptive* question of what the fixed method of science *is*, we would both uncover the logic of scientific change and answer the normative question of what method *ought to* be employed:

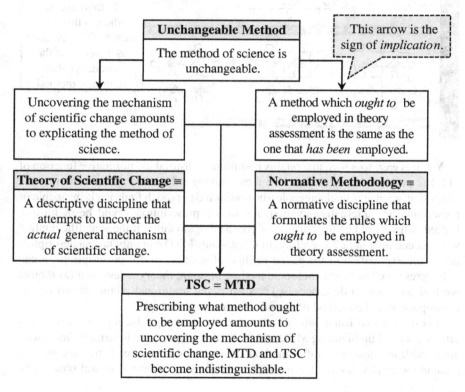

On this traditional view, therefore, there is no MTD separate from TSC; there is only one field of inquiry that is responsible for answering both normative and descriptive questions. Since the 1960s, this twofold descriptive-normative discipline has been assigned two inseparable functions – to uncover the logic of scientific change (which, in this reading, is the same as explicating the unchanging method of theory appraisal) and, thus, to openly "prescribe" this same method for future use. This traditional approach can be summed up thus:

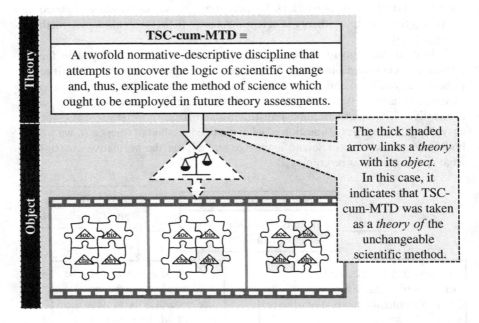

Yet, it is easy to notice that on this traditional construal the normative function of MTD is preserved only *de jure*. De facto, having subscribed to the view that the method of science is unchangeable, the traditional approach finds itself incapable of prescribing any alternative method, for such a prescription would be of no use. Indeed, what would be the purpose of prescribing anything, if the scientific method were indeed invariable? Therefore, this TSC-cum-MTD can only have a descriptive task – the explication of the actual method of science. As for prescriptions, it can only "prescribe" to follow that same method in future theory assessment (as though we had a chance to do otherwise).[26] Such is the traditional picture based on the assumption that the method of science is immune to change.

It is strange that Kuhn, who didn't accept the *unchangeable method thesis*, nevertheless ended up blending MTD and TSC. Nowadays, there is virtually incontestable evidence that scientific method is changeable. Clearly, the method of contemporary physics and the method of the sixteenth century natural philosophy

---

[26] Even those who deny the very possibility of any TSC cling to the tacit identification of the tasks of TSC and MTD. It is often argued that, since there are no fixed methods of science and since explicating this method is the sole task of TSC, there can be no TSC whatsoever (and, consequently, no MTD). I address this flawed line of reasoning in detail in section "The Argument from Changeability of Scientific Method" below.

are far from identical. But if we accept that the methods of appraisal are changeable, then it is obvious that methods must be treated as constituting part of the scientific mosaic and not something transcendent to it. This means that the evolution of the scientific mosaic involves not only changes in theories but also transitions from one employed method to another. Naturally, this turns the quest for the unchanging method of science into a pointless enterprise. Also, this transforms the question regarding the mechanism of scientific change, which now applies to changes in both theories and methods. How do the transitions from one accepted theory to another and from one employed method to another occur? In particular, how is one employed method being replaced by another? What laws (if any) govern this evolution? In short, as soon as we realize that the methods of appraisal are changeable, it becomes clear that *the laws of scientific change* and *methods* of appraisal are absolutely different entities:

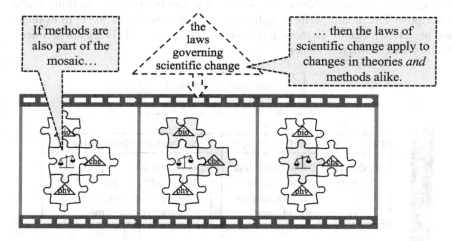

At the moment, we don't know whether scientific change obeys any general laws. What we do know is that these laws (if any) must necessarily apply to changes in theories and methods alike. Since TSC is a descriptive theory that must explain changes from one state of the scientific mosaic to another, we must concede that TSC should explain not only transitions from one accepted theory to another, but also from one employed method to another. It is the task of TSC to determine what laws (if any) govern the process of scientific change, i.e. the change of theories and methods.

Some authors have come to understand this as early as the 1980s. Larry Laudan, Dudley Shapere, and Ernan McMullin were among the first to point out that, since scientific change concerns not only theories but also methods, a theory of that process (TSC) should therefore attempt to explain transitions in theories and methods alike. Laudan's later theory, his famous reticulated model, is one such attempt.[27]

---

[27] See Shapere (1980), Laudan (1984), McMullin (1988).

What Laudan, Shapere, McMullin and others seem to have overlooked is that the picture of science with changing methods also implies that TSC and MTD no longer coincide. Since the task of TSC is to explain both changes in theories and changes in methods, it can no longer be expected to have any normative force. What its outcome could be has been clearly shown by Laudan himself in *Science and Values*: TSC should describe under what circumstances methods of science actually change. It may be that methods change because of the fundamental transitions in the ontology accepted by the scientific community, as both Kuhn and Laudan would have it. Or, it may turn out that transitions in methods are akin to changes in fashion, as Feyerabend would insist. In any case, the outcome of this investigation should necessarily be a descriptive theory. Thus, the only thing that we should expect from TSC is to describe and explain the actual process of scientific change. As for the normative task of *evaluating* methods and *prescribing* the best of them for future theory assessment, it would be too naïve to think that a descriptive TSC could perform it. It is the natural function of MTD to evaluate and prescribe methods. In short, it is evident that TSC and MTD are not identical, for they tackle different issues:

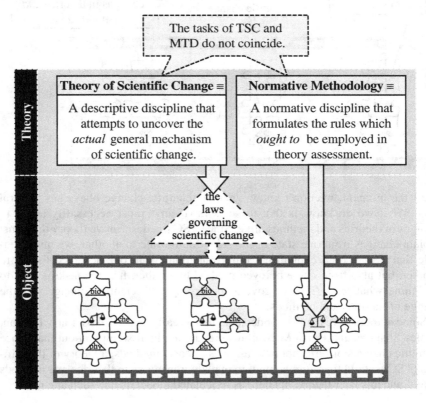

Keeping the two separate will save us from many mistakes, common among those who mix up TSC and MTD. For example, those who conflate the two often commit the following error. They take some method and evaluate it on the subject of its logical consistency and soundness and, if it turns out that the method in question has flaws, they proceed to claim that the method could not possibly be the one actually employed by scientists. It can be shown that this line of reasoning has the identification of TSC and MTD as its tacit premise. Indeed, if one equates TSC and MTD, one is forced to concede that only a sound and logically consistent method could contend to the title of the method of science. This follows from the tacit conviction that the normative question "what is the best available method?" and the descriptive question "what is the method actually employed in theory assessment?" should necessarily have *the same answer*. But this requirement could only be fulfilled if the actual method (provided that there *is* one fixed method) had no logical drawbacks.

Consider, for example, the case of the method of *induction*. It is a well-known fact that the method of induction (in all of its major formulations) has serious logical flaws such as the notorious *problem of induction* (the Hume problem) or the *paradox of confirmation*. From this, it has been often concluded that the method of induction could not possibly be employed by the scientific community in the actual theory assessment. The logical template is thus[28]:

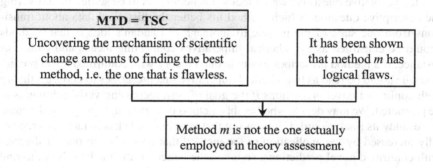

This same template has been used also to criticize the *hypothetico-deductive* method. It has been argued that, since the hypothetico-deductive method has flaws (e.g. the so-called *tacking by disjunction paradox*), it cannot possibly be employed by the scientific community in theory appraisal. Many Bayesian authors seem to submit to this line of reasoning.[29]

I believe that such a criticism is unfair, for it is based on a false premise. The fact that a method has some logical flaws doesn't imply that it has not been employed in actual theory assessment. There are no a priori grounds for claiming that all methods that have ever been employed were necessarily flawless; HSC seems to reveal quite the opposite. When we realize that MTD and TSC do not coincide, it becomes

---

[28] Both Popper and Lakatos employed this template. See Popper (1934/59) and Lakatos (1968).
[29] For discussion, see Nola and Sankey (2007, pp. 170–183).

obvious that a method might happen to be actually employed by the scientific community all its flaws notwithstanding. There is a straightforward analogy with scientific theories. Nobody claims that Newtonian physics was flawless and, yet, we all know that it was once the accepted physical theory. The same goes for any rejected theory that was once part of the scientific mosaic. And the same goes for any method that has ever been employed in theory assessment – being actually employed does not mean being flawless.

In short, we should not require descriptive TSC to evaluate the theories and methods which happen to be part of the scientific mosaic. Its task is to explain how changes from one state of the mosaic to another come about – changes in both theories and methods. The task of evaluating methods and, if necessary, prescribing them for future use pertains to normative MTD.[30]

Having shown that TSC and MTD do not coincide, we shall now address the question of their interaction – namely, the question of whether the findings of descriptive TSC and HSC have any bearing on normative MTD. When trying to evaluate what method is better (MTD), should we take into consideration our knowledge of how science actually works and how it has evolved (TSC and HSC)?

It has been suggested by Laudan among others that our evaluation of any given method should be based on its track record, on how it has actually performed during the history. It is one of the fundamental premises of so-called *normative naturalism* that the normative question "which method is better?" is to be settled by answering the descriptive question "which method fits better the historical data about transitions from one state of the mosaic to another?".[31] Laudan's idea is that methods should be construed as hypothetical imperatives connecting means and ends. For instance, if a method prescribes to single out a theory with confirmed novel predictions, it should be read as hypothetical imperative that prescribes to choose a theory with confirmed novel predictions *if* the goal of, say, increasing verisimilitude is to be promoted. We may decide whether this method is worth employing, says Laudan, if we study its track record. If it turns out that verisimilitude has in fact been repeatedly increased by the application of the method that prescribes to opt for theories with confirmed novel predictions, so much the better for the method. Now, to find out whether the method in question was or wasn't employed in a specific historical

---

[30] Garber is among few who see general TSC as purely *descriptive* theory that should merely *explain* the process of scientific change rather than *evaluate* it. See, for instance, Garber (1986), p. 94.

Nola and Sankey too seem to be realizing that it is not the task of descriptive TSC to evaluate scientific changes, but merely to *explain* their mechanism. However, they sometimes allow normative ingredients to enter from the backdoor. Consider how they formulate the following question: "what explains our choice of a sequence of theories which yield such instrumental success?" (Nola and Sankey 2000), p. 7. It is advisable, however, to refrain from such evaluative terms as "success". The question might be put in strictly descriptive terms: "what explains theory change?" full stop.

[31] The idea that historical record is to be used as a judge in methodological debates, did not originate with Laudan. It can be traced back at least to Kuhn, who expressed it on several occasions. He writes: "If I have a theory of how and why science works, it must necessarily have implications for the way in which scientists should behave if their enterprise is to flourish" (Kuhn 1970a), p. 237.

episode is a descriptive task which is normally assigned to HSC. Therefore, Laudan's claim amounts to saying that in order to fulfil the normative task of determining the best available method we should answer the descriptive question regarding the track record of available methods. Of course, the question remains how appealing this approach is; there is no unanimity whether it is justified to use historical data in judging what method is the best.[32]

Let us consider both options. If it turns out that MTD cannot legitimately use our knowledge about the workings of science, this will mean that TSC has no bearing on MTD. But even if it turns out that descriptive TSC could be legitimately used in settling the issues of normative MTD, it will change nothing in our understanding of the scope of TSC itself. For to say that MTD should base its prescriptions on the findings of TSC is analogous to saying that the engineer should draw heavily on existing physical theories when deciding what material to use for building a bridge. This does not, however, make physics and engineering identical. In particular, the choice of the proper material for bridge building does not become a task of physical theory. The same relation holds between TSC and MTD. In deciding what method is to be employed in theory assessment, normative MTD may draw upon our theory of how science operates (TSC, that is), but it doesn't make TSC and MTD identical.

Consider a similar analogy. The decision of a medical practitioner to prescribe some medicine in an instance of a certain illness is normally based on the knowledge of workings of the human body, provided by such a descriptive discipline as biology. The knowledge of human anatomy obviously has substantial bearing on the medical practice. This doesn't mean, however, that biology and medicine become indiscernible. The same goes for TSC and MTD: normative MTD may or may not base its prescriptions on the findings of descriptive TSC, but it doesn't make TSC itself a normative discipline.

Let us now sum up the outcome of this section. Theory of scientific change (TSC) should be understood as a descriptive discipline the main task of which is to explain the process of changes in the scientific mosaic. It is not identical with normative methodology (MTD), whose task it is to evaluate and prescribe methods. The findings of TSC may or may not be legitimately employed in such a normative evaluation, but TSC itself should not be expected to perform any normative functions: it is not to evaluate or prescribe anything.

Construction and Appraisal

The process of scientific change can be viewed from two different perspectives – from that of theory *construction* and that of theory *appraisal*. On the one hand, scientists can be viewed as constructing (generating, creating) new theories and thus

---

[32] There is vast literature on *normative naturalism*. Its outline may be found in Laudan (1996), pp. 125–179. For a shorter exposition, see Nola and Sankey (2007), pp. 321–329. For critical evaluation and discussion, see Doppelt (1990); Leplin (1990); Rosenberg (1990); Siegel (1990).

proposing modifications to the mosaic of accepted theories. On the other hand, scientists can be viewed as appraising proposed theories and making decisions as to which modification of the mosaic is to be accepted. Thus, one may ask two different descriptive questions:

Context of Discovery	**Context of Justification**
? How are theories *constructed*?	**?** How are theories *appraised*?

The objective of this section is to clarify which of these two questions TSC should address. Should TSC account for theory construction, or should it account for theory appraisal? In order to determine the answer, we shall start from the traditional distinction between *the context of discovery* and *the context of justification*.

Although it is Hans Reichenbach who is commonly considered the first to draw the distinction between the context of discovery and context of justification, it is widely recognized that the distinction had been implicit in the works of methodologists since the nineteenth century.[33] Popper, who clarified the difference between the two a few years prior to Reichenbach, traced it back to Kant's 1781 distinction between quid facti (questions of fact) and quid juris (questions of justification or validity).[34] The idea was simple: to draw a sharp demarcation line between the *temporal* process of generating (discovering, inventing, obtaining, constructing) theories, on the one hand, and the *logical* enterprise of justifying (appraising, testing, securing) the end product theory, on the other. It was the justification of fully developed theories that was considered a task of philosophy of science. As for the study of the actual process of generating (obtaining, constructing, discovering etc.) knowledge, it was taken as a task of such empirical disciplines as history, psychology, and the sociology of science. Implicit in this traditional distinction was the idea that, regarding the process of theory discovery, one can ask only *descriptive* questions, whereas the methods of justification are the subject of *normative* analysis.

It was also implicit in this distinction that the actual process of *discovery* could have no bearing upon the epistemological enterprise of *justification*. It was held by Reichenbach and Popper, among many others, that the provenance of a theory, the mode of its generation, the specific factors which led to its construction could play no role in its appraisal. As for the possibility of the logic of *discovery*, there was no unanimity among the proponents of this distinction. Some, like Popper, believed that there could be no such thing, since the process of discovery doesn't follow any logical algorithm and often contains an irrational element.[35] Others, like Reichenbach, considered it possible and even proposed special guidelines, which were supposed

---

[33] See Laudan (1980).

[34] See Popper (1934/59), p. 7. For discussion, see Peckhaus (2006).

[35] See Popper (1934/59), p. 8. It is ironic that the book that fervently denied the very existence of any logic of discovery was titled *The Logic of Scientific Discovery*.

to give heuristic advice for creating new theories and discovering natural regularities. What is essential, however, is that both parties agreed that even if there *were* a
logic of discovery, the justification of theories would remain unaffected by it, for in
reality the process of discovery and the process of justification are independent of
each other. Consequently, it was accepted that there are two completely different
questions – "how are theories being constructed (generated)?" and "how are theories justified (assessed)?".

This traditional distinction has been severely criticized ever since the times of
Hanson and Kuhn. One standard objection is that the process of discovery cannot be
viewed separately from that of justification. It has been argued that the two come in
an indistinguishable union, for knowledge is in fact both constructed and justified in
a single process. Thus, according to Feyerabend, "scientific practice does not contain two contexts moving side by side, it is a complicated mixture of procedures",[36]
and we should not, therefore, draw a demarcation line between the two contexts, for
there is actually one unified process. This argument has been repeated time and time
again. Friedrich Steinle, for instance, has recently written: "at the moment when
laws are formulated in the research process, they are discovered and justified at the
same time."[37]

It is nowadays generally agreed that the classical distinction between the two
contexts is in need of revision. In fact, there have been several attempts to modify
the classical distinction.[38] The latest (and perhaps the most successful) attempt is by
Paul Hoyningen-Huene, who proposed the so-called "*lean*" distinction between the
two contexts.[39] The basic idea is that there are two different *perspectives* that can be
taken regarding scientific knowledge. Compared with the classical distinction of
Reichenbach, Popper and others, Hoyningen-Huene's distinction leaves the question of the *actual* interrelation of discovery and justification aside. In particular, this
"lean" distinction does not presuppose that discovery and justification have no
actual bearing upon each other. It does not even presuppose that discovery and justification are two distinct processes: they may well turn out to be one indistinguishable process, just as Feyerabend and others have suggested. But that is irrelevant to
the distinction itself. What is stressed in the new distinction is that one may reasonably view the process from two different perspectives by asking two different *questions*. It is one thing to ask how scientific theories *are* generated (constructed) and it
is quite another thing to inquire how they *ought to* be appraised. Of course, it may
turn out that the two questions have the same answer, but that wouldn't make the
two questions identical.[40] It may turn out, for instance, that the only legitimate way

---

[36] Feyerabend (1975), p. 149. For discussion, see Hoyningen-Huene (2006), pp. 120–121.

[37] Steinle (2006), p. 187. See also Arabatzis (2006) from the same collection.

[38] Two important collections should be mentioned in this context: Nickles (ed.) (1980) and
Schickore and Steinle (eds.) (2006).

[39] For the distinction itself, see Hoyningen-Huene (2006, pp. 128–130). There are good indicators
of the success of this distinction in the same volume. See Schickore and Steinle (eds.) (2006), pp.
xiii, 134, 160, 188.

[40] See Hoyningen-Huene (2006), p. 129.

of appraising theories is by referring to the mode of their construction. Such a scenario is possible: the "lean" distinction itself does not impose any limitations in this respect. The only thing that the "lean" distinction suggests is that one may rightfully ask two different questions and that answering one question does not necessarily amount to answering another. For instance, it is one thing to ask how general relativity was constructed, and it is another thing to ask how it was appraised. It is conceivable that the process of its appraisal was tightly linked to the process of its construction, just as it may turn out that it wasn't. Therefore, the two questions may or may not have the same answer and that is all that the "lean" distinction assumes.

Obviously, the success of the "lean" distinction has to do with the fact that it doesn't demand too much, and rightly so. We can formulate it in the language of *object-* and *theory*-levels. Unlike the traditional distinction, the "lean" distinction refrains from saying anything about object-level processes of construction and appraisal and confines itself to stating the difference between the two questions, two perspectives. In other words, while the traditional distinction implied both the object-level independence of the *processes* of discovery and justification and the theory-level independence of two different *questions*, the "lean" distinction assumes only the theory-level distinction between two questions:

	Traditional distinction	"Lean" distinction
? Are the question of construction and the question of appraisal two different *questions*?	Yes	Yes
? Are theory construction and theory appraisal two essentially different *processes*?	Yes	*- no answer -*

The "lean" distinction is an apparent improvement to the traditional one. It is clear that, at this level – at the level of the metatheory – we should not commit ourselves to any specific position about the possible identity, interaction, or independence of the actual (object-level) processes of discovery and justification. The metatheory should only clarify which questions can be asked and which cannot. It is definitely not a task of the metatheory to substitute for a theory proper by providing answers to questions about object-level processes. Consider an example. Clarifying the scope of physics is a metatheoretical task. The respective metatheory – the metatheory of physics – should determine what questions physics should address and what questions it should not. But it is not a task of the metatheory to address questions about *actual* physical processes. For instance, the metatheory may prescribe that physics should account for such phenomena as solar or lunar eclipses. But the metatheory cannot and should not provide any actual explanations of eclipses, for that is a task for an actual physical theory. Similarly, our metatheory (the metatheory of TSC) should refrain from answering specific questions about the actual process of scientific change; it should rather clarify the list of issues which should be addressed by TSC, i.e. the scope of TSC. In this respect, Hoyningen-Huene's distinction is clearly superior to that of Reichenbach and Popper, for it avoids subscribing to any position regarding the actual relation between the processes of discovery and justification.

Nevertheless, although the "lean" distinction is obviously a step in the right direction, it shares with the traditional distinction one serious flaw. It is implicit in both distinctions that one may rightfully ask two questions: the *descriptive* question about theory generation (construction, discovery) and the *normative* question about theory appraisal (justification, assessment). The context distinction, on both accounts, virtually coincides with the distinction between descriptive and normative. Thus:

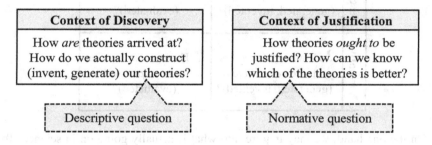

Context of Discovery	Context of Justification
How *are* theories arrived at? How do we actually construct (invent, generate) our theories?	How theories *ought to* be justified? How can we know which of the theories is better?
Descriptive question	Normative question

It is taken for granted that to study the process of discovery is to answer the *descriptive* question of how theories are *discovered*. Similarly, understanding theory *justification* amounts, on both accounts, to answering the *normative* question of how theories ought to be *appraised*. What both the traditional and "lean" distinctions do not realize is that there can be a *descriptive* question about *justification* as well as a *normative* question about *discovery*. Hence there can be not two but four different perspectives.

But before I propose my own fourfold distinction, I also find it is necessary to change labels. It has been noted by many authors that the choice of "discovery" and "justification" as labels for the two sides of the distinction is quite inappropriate.[41] There is enough evidence that calling the process of theory construction or generation "discovery" can be highly misleading. Sometimes authors forget that within this context distinction, "discovery" is understood in a special technical sense; instead, they use it in a sense which is perhaps intuitively more obvious, but is not the one employed in drawing the distinction. They use it, for instance, as in "the discovery of planet Neptune", or as in "the discovery of electron" or even as in "the discovery of the Zeeman effect". In any of these non-technical senses, "discovery" refers to an epistemic achievement, to something that has been positively appraised. The result of such an equivocation is that "discovery" becomes indistinguishable

---

[41] See, for instance, Leplin (1997), p. 86. Nola and Sankey (2007), p. 26.

from "justification".[42] I shall therefore refrain from using "discovery" and employ a more neutral "generation" or "construction" instead.[43] I shall also follow Lakatos and use "appraisal" instead of "justification", in order to avoid the unnecessary justificationist connotations of the latter.[44]

Thus, we can say that what both the traditional and "lean" distinctions do not realize is that there can be a *descriptive* question about *appraisal* as well as a *normative* question about *construction*. There are not two but four different questions:

	Construction	Appraisal
Descriptive	**?** How *are* theories actually *constructed* (generated, invented)?	**?** How *are* theories actually *appraised* (evaluated)?
Normative	**?** How *ought* theories to be *constructed* (generated, invented)?	**?** How *ought* theories to be *appraised* (evaluated)?

On the one hand, we may inquire into what is actually going on in science. To that end, we may formulate two *descriptive* questions – one focusing on theory *construction*, another focusing on theory *appraisal*. Answering the *descriptive* questions amounts to understanding the actual workings of science – how theories are in fact constructed and how they are in fact appraised. It requires studying the actual process (or processes) of theory construction and appraisal.[45] On the other hand, we may also pose two *normative* questions concerning *construction* and *appraisal*. In order to answer the *normative* question about theory *construction*, we would have to outline some heuristic guidelines that would prescribe how theories should be generated.[46] As for the *normative* question of *appraisal*, its answer is normally some set of criteria that a theory is expected to meet in order to become accepted.

It must be stressed that this fourfold distinction is meta-theoretical only. In this respect, it is similar to the "lean" distinction of Hoyningen-Huene, for it does not impose any limitations upon the actual interrelation of the four processes under study. It may well turn out that, say, the mode in which theories are constructed (a

---

[42] See Arabatzis (2006), p. 217.

[43] In this I follow Laudan (1980); Nickles (1992), p. 89; Sturm and Gigerenzer (2006).

[44] See Lakatos and Zahar (1976), p. 169. For discussion of the position of justificationism, see Lakatos (1970), pp. 10–11.

[45] Again, whether these two are identical, or interconnected, or completely unrelated is for actual research to uncover.

[46] See Langley et al. (eds.) (1987), especially pp. 37–62.

theory's provenance) plays an essential part in how they are actually appraised.[47] It may even turn out that the process of appraisal is the very process of its construction, as some have claimed.[48] A similar relation is conceivable between normative and descriptive perspectives. It could be that in order to understand how theories *ought to* be appraised one must study how theories *are* actually appraised, just as Laudan and others have insisted.[49] Similarly, it is conceivable that the best normative heuristics is in fact the one that is instinctively employed by scientists in actual theory construction. All these interrelations are, in principle, conceivable. It is vital, however, not to confuse the four different *questions*: a discipline that aims at answering one question should not end up answering the others.

Compared to both traditional and "lean" distinctions, the one I propose includes two additional questions – the *descriptive* question of *appraisal* and the *normative* question of *construction*. Yet, I am not the first to come up with these questions. The descriptive question of appraisal has been widely addressed since the times of Kuhn if not earlier. When philosophers of Kuhn's generation appreciated the importance of studying the actual workings of science, they basically set off to answer the *descriptive* question of *appraisal*. As for the discussions of the *normative* question of *construction*, they can be traced back (with some reservations of course) to the seventeenth century, if not earlier.[50] Since the mid-twentieth century, there have been numerous attempts to come up, as Herbert Simon puts it, with "a set of normative rules, heuristic in nature, that enhances the success of those who use them (as compared with those who don't) in making scientific discoveries".[51] Many heuristic problem-solving strategies have been since proposed.[52] The two "new" questions are, therefore, not exactly new.

But then a question arises as to how the normative question of discovery and the descriptive question of appraisal were situated within the twofold (traditional or "lean") distinctions. The short answer is that they were generally relegated to (subsumed under) one of the two contexts. Let us first consider the *descriptive* question of *appraisal*.

Some authors, like Reichenbach, would place it in the cell of the context of justification by welding it with the *normative* question of *appraisal*. This move was quite natural for those who believed in the fixed method of science, as I have shown in section "Descriptive and Normative". Indeed, those who hold that there is an unchangeable method of science are naturally inclined to think that answering the *normative* question of how theories *ought to* be appraised amounts to answering the descriptive question of how theories *are* actually appraised. The method that ought

---

[47] See Leplin (1997), pp. 56, 58, 66.

[48] See Schickore and Steinle (eds.) (2006) for discussion.

[49] This is the main premise of Laudan's *normative naturalism*. See Laudan (1996), pp. 134–138.

[50] See Koertge (1980).

[51] Simon (1992), p. 75.

[52] See Nola and Sankey (2007), pp. 22–28 for an overview and references.

to be employed in theory appraisal, on this reading, is identical with the method that has been actually employed in theory appraisal.[53]

Others, like Lakatos, would consider the descriptive question of appraisal as pertaining to the context of discovery and, thus, weld it with the *descriptive* question of *construction*. Like many others, Lakatos thinks that "explanation of change (of actual acceptance and rejection of theories) is a psychological problem".[54] It is, he believes, the same as "the problem of why and how new theories emerge".[55] Thus he equates two different descriptive issues – the question of construction (how new theories emerge) and the question of appraisal (how theories actually become accepted). This identification of the two descriptive issues is widespread even nowadays. Sturm and Gigerenzer, for instance, commit the same mistake when they write: "For any given claim p, we can ask, 'How did someone come to accept that p?' This question, which may be understood as a question about the generation or actual acceptance of a claim, differs in principle from the question 'Is p justified?'".[56] Now, they are correct when they say that the descriptive question of how p was actually accepted is *not* the same as the normative question of how p ought to be appraised ("justified"). But to say that the descriptive question of how p was actually *accepted* may be understood as a question about *generation* (!) of p is a gross distortion of the picture. Sturm and Gigerenzer simply blend the two descriptive questions – that of construction and that of appraisal.

Conflations like these are inevitable if one subscribes to any twofold context distinction, for one has to position the descriptive question of appraisal somewhere, whereas the twofold distinctions provide only two possible slots.

The situation with the *normative* question of *construction* is no better. Under the common heading of the logic of discovery it has been often mixed up with the descriptive question of construction. The normative task of providing a heuristic guidance for theory construction has been conflated with the descriptive task of studying the actual process of theory construction. They were both traditionally considered pertaining to the context of discovery. This is surprising indeed, for, when forced, nobody would equate the task of descriptive cognitive psychology with that of formulating a set of normative heuristic rules that should guide us in theory construction.[57]

Hence, to avoid such confusions, we must differentiate between not two but *four* perspectives, instantiated by the four questions. It is vital for our purpose not to conflate them in the future.

Following this lengthy preliminary spadework, we can now formulate the main meta-theoretical problem of this section: which of these four questions should TSC deal with? Namely, should TSC address the descriptive question of construction and

---

[53] Reichenbach even utilizes a special "camouflage-terminology" that conceals the conflation of the two. See Nickles (1980), p. 11 and footnote 21 on p. 51.

[54] Lakatos and Zahar (1976), pp. 168–169.

[55] Ibid.

[56] Sturm and Gigerenzer (2006), p. 134.

[57] See Nickles (1980); Simon (1992); Nola and Sankey (2007), pp. 22–25.

explain how theories are actually constructed? Should it address the descriptive question of actual theory appraisal? Should it prescribe heuristic rules for theory construction? Or should it prescribe methods for theory appraisal?

Let us start with the two *normative* questions – that of construction and that of appraisal. As I have shown in section "Descriptive and Normative", TSC is a *descriptive* theory and is not immediately charged with the duty of addressing normative issues. It was implicit in my earlier discussion that the normative question of *appraisal* pertains to the realm of normative methodology (MTD). It is the task of MTD, not TSC, to prescribe what methods of theory assessment ought to be employed. As for the normative question of *construction*, being once considered a proper question for normative MTD, it was then expelled from the domain of MTD and is nowadays tackled by several interrelated fields, such as, artificial intelligence and normative decision-making.[58] There are authors nowadays who argue that the normative question of construction should be brought back to the realm of MTD, where it once belonged.[59] In any case, what is relevant from our perspective is that both of these normative issues are unquestionably beyond the scope of descriptive TSC.

Thus, we are left with the two *descriptive* questions. While the case of the descriptive question of *appraisal* is relatively unproblematic, that of the descriptive question of *construction* is challenging. It is obvious that TSC must account for the actual process of theory appraisal. By definition, TSC should describe and explain how changes in the mosaic of accepted scientific theories and methods take place. Any actual instance of scientific change is nothing but a result of an appraisal – a decision of the community to accept a proposed modification to the mosaic. Therefore, TSC must provide an account of how theories (or proposed modifications to the mosaic, to be more precise) are actually appraised and, by that, explain how transitions from one state of the mosaic to another actually take place.

What is not so clear is whether TSC should also answer the other descriptive question – that of construction. The definition of TSC is silent in this respect. Indeed, it says nothing explicit about theory construction. Hence, it is safe to say that TSC should not necessarily concern itself with theory construction. We can state it more precisely: the process of theory construction may become relevant to TSC if it turns out that it has some bearing upon the process of theory appraisal. Whether there is such a bearing is irrelevant at this point as it is for an actual TSC to settle this issue. What is essential from the metatheoretical perspective is that TSC *is not required* to account for theory construction.[60]

---

[58] For a historical excursion, see Laudan (1980).

[59] See Nickles (1980).

[60] We should also remember that nowadays there are already several disciplines directly concerned with this question. The descriptive question of theory construction is tackled by such disciplines as descriptive psychology and sociology of science, so the need of another discipline attempting to illuminate the process of theory construction is questionable. It is therefore reasonable to conclude that TSC should try to refrain from duplicating the labours of these disciplines. At the same time, if it turns out that the provenance of the theory plays a role in its appraisal, TSC can use sociologi-

Let us recap the main outcome of this section. Having considered both the traditional and "lean" distinctions between discovery and justification, I have proposed a more suitable fourfold distinction with four different questions. It is evident that, of these four questions, only the descriptive question of appraisal is strictly speaking within the scope of TSC. On the other hand, TSC is allowed to *refer to* the answers to the three other questions, provided that TSC needs those answers in order to fulfil its main task – the task of explain the process of scientific change. The latter is a necessary condition: TSC may employ the findings of cognitive science, sociology, psychology, or any other discipline *only insofar as* that is required for fulfilling the task of explaining theory appraisal. Likewise it may turn out that the findings of TSC have consequences important for methodology, psychology, sociology, or some other discipline, but that can only be a *by-product* of TSC, not its main purpose.

Acceptance, Use, and Pursuit

So far we have clarified that it is the task of TSC to explain actual transitions from one accepted theory to the next and one employed method to the next. But what do we mean when we say that a theory is *accepted*? Unfortunately, historians and theoreticians of science often talk about "acceptance" without clarifying in what sense a theory is said to be "accepted". Is a theory accepted when the scientists involved in the field declare that they believe it is true, or is it accepted when they are actually involved in its elaboration? Or is it to be considered accepted when it is used in calculating predictions of future events or when it becomes a basis for technology and is used in, say, bridge building? Is there a difference between developing (elaborating, modifying) a theory and appraising it? To answer these questions, we have to clarify the difference between *acceptance*, *use*, and *pursuit*. The clarification of these terms is a matter of urgency since the confusion of any two of them leads to a serious misunderstanding. Many different words have been used to describe attitudes that the scientific community can possibly take towards a theory, mostly without any attempt to clarify their respective meanings. Kuhn alone used a number of different and equally vague words, such as "universally received", "embraced", "acknowledged", "committed".[61] It is not evident whether these terms are meant as synonyms or whether there are important differences between them. So when the historian says that some theory was universally received, the exact attitude of the community towards that theory often remains a mystery. Thus, a clear and unambiguous nomenclature of the possible stances that the scientific community can take towards a theory is a must.

---

cal and psychological theories about theory construction for the purpose of explaining theory appraisal. See, for instance, Leplin (1997), pp. 48–58.

[61] See Kuhn (1962/70), pp. 10–13.

I propose to distinguish between the following three stances:

Acceptance ≡	Use ≡	Pursuit ≡
A theory is said to be accepted if it is taken as the best available description of its object.	A theory is said to be used if it is taken as an adequate tool for practical application.	A theory is said to be pursued if it is considered worthy of further development.

When we say "a theory is accepted" we will mean that it is considered the best description of its object we have at hand. When scientists accept a theory, they may clearly understand that it is not true in the strictest sense. While historically there have been scientific communities that believed that their mosaics contained infallible truths about the universe, from a logical standpoint it is not necessary to believe that a theory is *absolutely* true in order to accept it. Suffice it to believe that, compared to its rivals, the theory in question provides the best available description of its object, whatever that object is. Thus, nowadays, we accept *general relativity* in the sense that we think it provides the best extant description of the processes of its domain.[62] Similarly, we accept the *modern evolutionary synthesis* in the sense that we consider it as the best available description of the process of biological evolution. The same goes for any other *accepted* theory.

Acceptance in this specific technical sense should be differentiated from what has often been called "instrumental acceptance", or what I shall call *use*. In order to be considered *useful*, a theory does not need to be taken as the best available description of its object. A theory is said to be *useful* if it is considered valuable in practical applications, such as, say, constructing microchips, crop yield increase, or election winning. Importantly, a *used* theory may or may not be accepted by the community as the best extant description of its object. Quite often, we accept one theory but use another theory which we do not accept. Take an example of classical physics which is no longer considered as the best available description of its domain. In fact, it was replaced in the mosaic during the 1920s by general relativity and quantum physics. Yet, when it comes to bridge building, our engineers do not use the equations of general relativity (unless, of course, it is a transcosmic bridge); instead, they still use the equations of the good old classical physics, for its equations are pretty good for the task at hand. Thus, albeit no longer accepted, classical physics is still *used* in many practical applications.

---

[62] There is a discussion in the literature on whether one may legitimately apply "belief" to groups (i.e. "a group believes that p") or whether one should stick to a more traditional "acceptance" ("a group accepts p"). See Clarke (1994); Wray (2001). In order to avoid unnecessary complications, I shall stick to "acceptance". Whether our linguistic intuition allows for such phrases as "community believes that p" is irrelevant in this context. What is relevant is that when we say "the theory is accepted by the community" we mean that the community considers the theory in question as the best description we have at hand.

Finally, *acceptance* and *use* are not to be confused with *pursuit*. A theory is *pursued* when its advancement is considered promising in one sense or another. Sometimes scientists think that an individual idea or even a full-fledged theory may be developed in some interesting way and, thus, decide to work on elaborating it. It is not necessary for such a theory to be accepted: scientists should not believe that the theory they're working on provides the best existing description of its object. One may devote oneself to *pursuing* some idea without accepting it as the most correct description at hand. Nor must a theory be useful in order to be pursued. It is sufficient if a theory is considered worth developing. For example, many physicists currently work on elaborating and advancing different superstring theories, while none of these theories is either currently accepted or particularly useful. The opposite is also true: one may believe that a theory provides the best available description of its object without committing oneself to working in that direction. In short, it is possible to *accept* one theory, to *use* another theory in practice and, at the same time, to *pursue* some other promising theory.

Naturally, all these stances come with their negations. The opposite of *used* is *unused*: when scientists think that a theory is not convenient in some specific respect, we say the theory is considered *useless* in that respect, i.e. it is *unused*. The opposite stance of *pursued* is *neglected*: when scientists do not think that a theory is worth developing, when they do not attempt to advance (elaborate) it, we may say that a theory is *neglected*. Finally, the opposite of *accepted* is *unaccepted*: when scientists do not consider a theory as the best available description, we say the theory is *unaccepted*. *Unaccepted* should not be confused with *rejected*. *Rejection* is a type of scientific change, where an element of the scientific mosaic (a theory or a method) *ceases to be* part of the mosaic. A theory may always remain *unaccepted*, while in order to be rejected, a theory needs to be previously *accepted*. For instance, the so-called M-theory (a version of string theory) is currently unaccepted, but it has never been rejected, since it has never been accepted in the first place. Cartesian physics, on the other hand, was rejected in the mid-eighteenth century and is currently unaccepted.

	Yes	No
? Is a theory taken as the best available description of its object?	Accepted	Unaccepted
? Is a theory employed in practical applications as a useful tool?	Used	Unused
? Is a theory considered worthy of further development, elaboration?	Pursued	Neglected

Let us start from the distinction between *acceptance* and *pursuit*. Although it might be tempting to trace this distinction back to David Hume's differentiation between *believing* and *entertaining*,[63] the difference between *acceptance* and *pursuit* was first clarified by Laudan. As he explains in his *Progress and its Problems*, to work on modifying (elaborating, advancing) a theory does not necessarily mean to accept it.[64] He is right to point out that neither Kuhn nor Feyerabend see the difference between the two.[65] Feyerabend, for instance, often uses such confusing phrases as "the theory becomes acceptable as a topic for discussion".[66] Clearly, here "acceptable" is taken as a synonym for "pursuit-worthy".

Laudan, however, was not the first to utilize this distinction. It is safe to say that Lakatos tacitly assumed a similar distinction in his *Falsification and the Methodology of Scientific Research Programmes*. On the one hand, Lakatos proposes his explication of the scientific method – his methodology, which includes his famous three rules and his distinction of progressive and degenerative research programmes.[67] On the other hand, he admits that scientists may rationally hold on to their favourite research programmes for as long as their ingenuity allows. Some interpreters, such as Feyerabend, have accused Lakatos of self-contradiction. They read Lakatos as saying that we should *accept* the best of the available competitors but, at the same time, we are free to *accept* whatever theory we please.[68] Clearly, this reading of Lakatos is erroneous. He himself deflects the criticism by saying that the seeming contradiction vanishes once we differentiate *appraisal* (to accept or not accept)

---

[63] See Hume (1739/40), p. 83.

[64] See Laudan (1977), pp. 108–114. At times, however, Laudan seems to be forgetting about this distinction. Take for instance his criticism of Kuhn's mono-paradigmatic view of normal science, when Laudan points out that there is normally more than one paradigm at any period of history. Naturally, Laudan cannot mean that two or more paradigms may be simultaneously *accepted* – they can be only simultaneously *pursued*. But this weaker thesis would not hurt Kuhn, for Kuhn's position can be reconciled with the view that, at any given period of time, there is only one *accepted* paradigm, but many *pursued* paradigms. See Laudan (1977), pp. 134–137.

Similarly, when he claims that "those scientists who take up a theory in its early stages subscribe to different standards from those who take it up, if at all, only at a much later stage in its development", he tacitly assumes that *pursuit* and *acceptance* are the same stances. Laudan and Laudan (1989), p. 224.

[65] See Laudan (1984), pp. 14, 16.

[66] Feyerabend (1975), p. 30; cf. pp. 14, 24. For other examples of the conflation, see Feyerabend (1981), p. 105; Kuhn (1977), p. 332; Newton-Smith (1981), p. 231. The following quote from van Fraassen is illustrative: "to accept one theory rather than another involves also a commitment to a research programme, to continuing the dialogue with nature in the framework of one conceptual scheme rather than another" (van Fraassen 1980), p. 4. Cf. also p. 88. Even nowadays, some authors do not always differentiate the two. See, for instance, Nola and Sankey (2000), pp. 12, 31; Kieseppä (2000), p. 341.

Among those who do recognize the difference are Brown and Whitt. See Brown (2001), pp. 90–91; Whitt (1990).

[67] Lakatos's rules are, in a sense, elaborations of Popper's three rules. Two of Popper's rules first appear in Popper (1934/59), pp. 61–63. The third rule, suggested by Agassi, is presented in Popper (1963), pp. 326–336. Lakatos introduces his rules in Lakatos (1970), pp. 32–34.

[68] See Feyerabend (1970), pp. 215–216.

from *heuristic advice* (to pursue or neglect). When it comes to *appraising* the available competitors, there are, according to Lakatos, certain criteria that determine which of them has the best track record. But this doesn't apply to the question of how we determine the most *promising* course of research. Lakatos insists that we should not impose any limitations on which of the competitors are worth *pursuing* (developing, elaborating). First of all, it is difficult to tell from the outset which initial idea is capable of growing into a full-fledged theory.[69] In addition, history knows many examples when long-degenerating programmes suddenly make glorious come-backs (e.g. atomism). In short, Lakatos's position is that we should allow different competing ideas to be elaborated, but it doesn't mean that we shouldn't keep score. If we use my terminology, this amounts to saying that there is freedom in choosing which idea to *pursue*, elaborate, but the results must be judged and the best current option should be *accepted*.[70]

The difference between *acceptance* and *pursuit* was also emphasized by Stephen Wykstra. Insofar as I know, he was also the first among philosophers to openly distinguish between *acceptance* and *use* as early as 1980.[71] He emphasizes the necessity of resisting talk about "commitments" or "acceptance" in a vague sense and proposes to differentiate several "cognitive stances" that can be taken toward theories. "To commit oneself to working on a theory is one sort of cognitive stance; to take the theory for granted in testing other theories is another; ... and to use the theory to put men on the moon, yet something else." Compared to Laudan, Wykstra takes a step forward, for Laudan doesn't see the difference between *acceptance* and *use*.[72]

In this, Laudan is not alone. Among those who confuse what I call *acceptance* and *use* is Bas van Fraassen, who thinks that "the belief involved in accepting a scientific theory is only that it 'saves the phenomena', that is correctly describes what is observable."[73] Van Fraassen clearly conflates *acceptance* and *use*. On his definition, in order to accept a theory it is necessary to believe that it is the best calculating tool there is, that is to "rely on it to predict the weather or build a bridge".[74] There is no need, says van Fraassen, in believing that it is the best description of reality. In van Fraassen's case, this conflation comes as no surprise, for according to his antirealist epistemology one cannot legitimately believe that a theory is true about the world (if "the world" also includes what is unobservable).

---

[69] The way he presents it is that we should not "conflate methodological appraisal of a programme with firm heuristic advice about what to do" Lakatos (1971), p. 117. Recently, Nickles has used similar wording to stress the difference between acceptance and pursuit. In Nickles's language, it corresponds to the distinction between epistemic appraisal and heuristic appraisal. See Nickles (2006).

[70] See Lakatos (1971), p. 117. For discussion, see Motterlini (1999), pp. 4–5.

[71] See Wykstra (1980), p. 216. The distinction between *use* and *acceptance* is also implicit in Mario Bunge's *Treatise*. See Bunge (1983), p. 114.

[72] Interestingly, virtually all Laudan's examples of acceptance are cases of what I call *use*. See, for example, Laudan (1977), pp. 108–109.

[73] Van Fraassen (1980), p. 4.

[74] Van Fraassen (1980), p. 151.

That is why, for van Fraassen, "the only belief involved in accepting a scientific theory is belief that it is empirically adequate".[75] "To accept the theory involves no more belief, therefore, than what it says about observable phenomena is correct."[76] If we were to follow van Fraassen, we would end up conflating *acceptance* and *use*.

The same holds in the case of Nancy Cartwright. Although not quite in these terms, she too seems to equate *acceptance* and *use* when she says "we all know that quantum physics has in no way replaced classical physics. We use both".[77] Similar to Van Fraassen, she does this deliberately, for she holds that accepting a theory amounts to taking it as practically useful. This brings Cartwright to her "patchwork" view of science, according to which, at any given time there are many different laws that are unrelated to each other in any systematic way; which of these laws "we choose from one occasion to another depends on the kinds of problems we are trying to solve".[78] In short, the stance that I call *acceptance* is absent in Cartwright's nomenclature.

Now, whether one can *legitimately* believe in anything with regard to what is unobservable, is an *epistemological* question of crucial importance – a question that demarcates realists and antirealists.[79] Yet, albeit essential for epistemology, this issue is completely irrelevant to our task. Our question is not epistemological, but factual: what different stances can the scientific community *actually* take towards a theory? What is important from the perspective of TSC is to have a meaningful classification of the stances that the scientific community can *actually* take towards theories. It is a historical fact that some theories have been *accepted* by the community as correct descriptions of their respective domains, others have been treated as *useful* calculating devices, while still others have been both *accepted* and *used*. For this reason alone we have to distinguish between *acceptance* and *use*. Nowadays, for instance, it is accepted by the scientific community that *quantum optics* is the best available description of its domain. Yet, this does not discourage us from using the good-old *classical optics* in all sorts of practical applications (e.g. in calculating atmospheric refraction, building telescopes etc.). Again, it is an important epistemological issue whether all or any of these stances of the scientific community are epistemologically valid; it is interesting to know whether scientists are justified in differentiating between *acceptance* and *use*. However, the outcome is irrelevant to our purposes, for all we want to know in this context is actual stances of the community. Historically, some theories have been considered as providing better descriptions, while others have been taken as useful tools. This historical fact is all that we are interested in here. The question of the legitimacy of these stances is extremely interesting, but irrelevant: it is not the task of descriptive TSC to evaluate

---

[75] Van Fraassen (1980), p. 197.

[76] Van Fraassen (1980), p. 57. For critical discussion see Horwich (1991).

[77] Cartwright (1999), p. 2.

[78] Ibid.

[79] For a thorough discussion of the issue see Chakravartty (2007) and references therein.

or prescribe, but to describe and explain.[80] It is this point that both Van Fraassen and Cartwright seem to have missed.

Cartwright's patchwork view, however, has illuminated one important difference between *acceptance* and *use*. If we use my terminology, Cartwright may be taken as saying that new useful theories do not necessarily replace the old useful theories. With this I can only agree: as far as the context of *use* is concerned, what we have at any given point of time is a *toolbox*[81] of many different theories that provide different tools for different practical purposes. Thus, when a new hammer is added to a toolbox, the old hammer must not necessarily be thrown away; they both can be used depending on the task at hand. Similarly, an old used theory need not necessarily be thrown away when a new useful theory is found. Both of these theories can be used depending on the task.

But there is more to the story than Cartwright allows. One crucial difference between *acceptance* and *use* is that, whereas *useful* theories of the toolbox may accumulate without replacing the old theories, the newly accepted theories normally supersede some previously accepted theories. Say there are two mutually incompatible astronomical theories. One of the theories provides the most precise and accurate predictions of the positions of Mars but lacks, at the same time, the precision and accuracy of the other astronomical theory in predicting the positions of, say, Jupiter. Obviously, these two theories cannot be simultaneously accepted (since they are incompatible), but they can still be simultaneously used – the theory which is better in predicting the trajectory of Mars will be more useful in that particular application, while the other theory will be more useful for predicting the future positions of Jupiter. Similarly, a physical theory successfully used in, say, building frigates may be incompatible with a physical theory successfully used in spaceship construction. Yet, only one of these theories may be considered as providing the best available description (that is, accepted), whereas both of these theories can be simultaneously used in practice if they turn out to be useful. It is safe to say that, from the practical standpoint, there is an ongoing process of accumulation of tools in the toolbox, where many old theory-tools continue to be used alongside new theory-tools.[82]

It must be noted that some historians drew the distinction between *acceptance* and *use* even before Wykstra. Namely, Robert Westman pointed out that we often assign different meanings to "acceptance". "Acceptance may connote provisional use of certain hypotheses (without commitment to truth content), or acceptance of certain parts of the theory as true while rejecting other propositions as false, ... or acceptance of the theory as true without regarding it as a program for further

---

[80] See section "Descriptive and Normative" above for discussion.

[81] *Toolbox*, I believe, is a better term than *patchwork*, for the latter might be easily confused with the mosaic of accepted theories. And that's all we need – yet another confusion of terms!

[82] When d'Espagnat argues that science is cumulative, what he has in mind is precisely this sort of accumulation of new calculating tools. See d'Espagnat (2008), p. 146.

research."[83] It is my impression that many historians understand the need for such a differentiation,[84] but the absence of a uniform terminology often results in clumsy attempts to clarify the key notions within individual historical narratives. As a result, new terms such as "adopted" or "sustained" are being introduced, which leads to a further proliferation of vague terms.[85] That is why there is an urgent need for an unambiguous nomenclature that would clearly distinguish between *acceptance*, *use*, and *pursuit*.

In order to illustrate the proposed distinction between *acceptance*, *use*, and *pursuit*, let us consider some historical examples. It is well known that the *Aristotelian-scholastic cosmology* was accepted during the late Middle Ages. According to this theory, the heavens consisted of a set of tightly nested concentric spheres, the complex motion of which roughly described the apparent motion of the planets. There was also the *Ptolemaic astronomical theory* which described the apparent motions of the celestial bodies by means of a complex system of eccentrics, deferents, and epicycles. As far as the accuracy of predictions was concerned, the Ptolemaic theory was greatly superior to the Aristotelian cosmology, which was at best capable of providing general qualitative predictions. The trouble was, however, that initially the Ptolemaic astronomy was considered incompatible with the Aristotelian cosmology. The eccentrics and epicycles of the Ptolemaic astronomy were not readily reconcilable with the concentric spheres of the Aristotelian cosmology. Reconciling the two theories was a serious challenge for many generations of medieval astronomers. One such reconciliation was given by Ibn al-Haytham, who suggested that each planetary sphere was thick enough to contain in itself an eccentric channel through which the ring of the epicycle passed, through which, in turn, passed the planet. Albeit somewhat awkward, this "patch" managed to reconcile the Ptolemaic theory with the then-accepted Aristotelian theory of concentric spheres. After the reconciliation was provided, the scientific community could accept both the Aristotelian and Ptolemaic theories (together with al-Haytham's "patch").[86]

A question arises, however: what was the stance of the community towards these two theories *before* the reconciliation. Tradition has it that that the Ptolemaic astronomy was accepted together with the Aristotelian cosmology even before they were

---

[83] Westman (1975), p. 165. Naturally, it should not puzzle us that Westman does not use such words as "use" or "pursuit". What is important is the distinction of the *concepts* that he makes, not the *words* he uses.

[84] Henry Guerlac is one example. In his account of the reception of the Newtonian theory in France before the 1730s, he finds it necessary to clarify that whereas Newton's laws were considered empirically adequate tools (i.e. *used* in my terminology), they were not taken as describing reality (i.e. they were *unaccepted*), for such notions as *void* and *attraction of bodies at a distance* were seen as apparently absurd. See Guerlac (1981), p. 62. Peter Dear is yet another example. Although he doesn't use precisely the same language, he certainly sees the difference between accepting a theory as the best available description of its object and using it as a mere calculating tool. See Dear (2005), pp. 403–404.

[85] See, for instance, Wilson (1989a), pp. 161–162; Friedman (2001), p. 22.

[86] See Lindberg (2008), pp. 261–270. There are indications that a version of al-Haytham's reconciliation was accepted in Paris by early seventeenth century. See Ariew (1992), p. 358.

reconciled. But how could they both be simultaneously accepted if they were mutually incompatible? It would be all but impossible to answer this question without a strict distinction between *acceptance* and *use*. With these two stances available, the solution follows naturally: during the period when the Ptolemaic astronomy was thought to be incompatible with the then-accepted Aristotelian cosmology, it was merely used, i.e. it was taken as a good calculating tool, useful in practical applications (e.g. horology, astrology and, consequently, medicine). During that period, the scientific community couldn't and didn't take the eccentrics and epicycles for real. However, once the reconciliation was available, the Ptolemaic model could become accepted (without losing its status of the most useful theory, of course).

Another example is provided by Westman. In his account of the early reception of the Copernican theory, he correctly pointed out that although the theory was *unaccepted* (to use my terminology), some of its elements were widely *used* at the University of Wittenberg as tools for calculating the positions of celestial bodies. Westman calls this the *Wittenberg interpretation* of the Copernican theory.[87] Naturally, the accepted theory was a version of the Ptolemaic astronomy (with all its medieval and early modern modifications). Of course, the situation had changed after the geocentric view of the world was replaced by the conception of heliocentric solar system in an infinite (and, thus, centerless) universe.[88] What is important here, and what Westman convincingly shows, is that there is no conflict in holding that a theory can be widely used without being accepted.

These days we have an analogous case. The accepted theory that supposedly describes the universe on a large scale is *general relativity*, proposed by Einstein and developed by several generations of scientists into its current state.[89] It is a textbook fact that general relativity came to replace the Newtonian theory of gravity as the accepted theory. However, in an immense number of practical applications we still use the good old classical theory of gravity, although we no longer consider it the best available description of its domain. In particular, we no longer accept a force of gravity acting at a distance through empty space; instead we ascribe attraction effects to spacetime curvature. Had we not made a distinction between *acceptance* and *use*, we would have been forced to deal with a puzzling situation, where two mutually incompatible theories were simultaneously "accepted".[90] Consequently, we would have been forced to subscribe to something similar to Cartwright's patchwork view. That is why, it is important to appreciate that *acceptance* and *use* are different stances and that they are not to be confused.

Cases like these are innumerable. They make the distinction between *acceptance* and *use* so obvious that it is really surprising that some theoreticians and historians still manage to confuse the two. Perhaps what complicates the situation is that, even within the same theory, some propositions may be accepted, some only used, and

---

[87] This somewhat cumbersome language can be readily excused, given that in Westman's times there was no special terminology for the treatment of such cases.

[88] See Koyré (1957).

[89] See Zahar (1989); Penrose (2004), pp. 440–470, 686–734.

[90] Peter Dear has realized this. See his (2005), p. 403, footnote 36.

some both accepted and used. As Westman has indicated, it is quite possible to accept only some parts of a theory without accepting every word of it.[91] The current stance of the scientific community towards quantum physics seems to be a good example. When it comes to the question of the status of quantum physics, it is often noted that the scientific community takes it as an extremely useful calculating tool but doesn't accept it (in the technical sense of the term). Although there is some truth in this statement, it is not exact. On closer scrutiny we can observe that, while some of its propositions are unaccepted but merely used (e.g. *the collapse postulate*), other propositions are not only used but also accepted (e.g. *quantum nonlocality*). Take, for instance, the Standard Model of quantum physics which is not only used but also clearly accepted as the best description of the domain of elementary particles. Nowadays, scientists do believe that there are in fact leptons and quarks, that they are fundamental constituents of matter characterized by their respective properties and governed by their respective quantum laws. The recent discovery of the Higgs boson was praised not because of its extreme usefulness, but because it added a new particle to our accepted list of fundamental particles. So it is safe to say that a considerable portion of quantum physics is clearly accepted. Therefore, one cannot agree with Arthur Fine's characterization of quantum physics as "the blackest of black-box theories; a marvellous predictor but an incompetent explainer".[92] It is not quite "black", for at least some of its propositions are taken nowadays as best available descriptions of the microworld.

From the distinction between *acceptance* and *use*, let us now turn to the distinction between *acceptance* and *pursuit*. I shall consider some historical examples which will help to illustrate it.

My first example is from the mid-seventeenth century. Despite what we encounter in both the popular and professional literature on the so-called Scientific Revolution, the Aristotelian-scholastic natural philosophy, with its theory of elements and four causes, its geocentric cosmology, its Aristotelian laws of motion as well as many other constituents remained accepted up until the end of the seventeenth century.[93] Where our textbooks and encyclopaedias are correct is in emphasizing that many other directions were pursued at that time. Many of the natural philosophers of the seventeenth century, whose names we nowadays include in our encyclopaedias and textbooks, were pursuing one or another direction with the aim of overthrowing the accepted Aristotelian-scholastic natural philosophy. There was, of course, Galileo with his two new sciences. There was also the *mechanical natural philosophy* in all its different versions (Beeckman, Descartes, Huygens, and Boyle, among many others).[94] In addition, there were those who worked on the *magnetical*

---

[91] See Westman (1975), p. 165.

[92] Fine (1982), p. 740.

[93] See Brockliss (2003), pp. 45–46. One cannot agree with Heinrich Kuhn when he says that "[i]t seems difficult (or even impossible) to find a single statement on which all known Renaissance 'Aristotelians' agree" (Kuhn 2005). This is clearly an exaggeration, since many tenets of the Aristotelian-medieval natural philosophy were universally accepted.

[94] See Boas (1952).

philosophy (Gilbert, Kepler, Stevin, Wilkins, to name only a few).[95] Others, like Gassendi or Newton, were pursuing yet other directions. However, some exceptions aside, for most of the seventeenth century these directions were only pursued but unaccepted.[96] Those historians who confuse the two, end up enunciating that the Scientific Revolution started around 1543 with the publication of *De Revolutionibus* and culminated in 1687 with Newton's *Principia* – a gross misconception as far as the history of the scientific mosaic is concerned.[97]

Laudan provides another example from the early 1800s. According to the then-accepted chemical theory, chemical species had the tendency to combine with certain substances or species in preference to others. It was believed that analogous substances showed the varying degrees of affinity for different reagents. On the contrary, Dalton's atomistic chemistry, proposed in the early 1800s, was an attempt to explain seeming affinities between different chemical substances by providing an account of the constituents of matter. To be sure, Dalton's own theory in its original formulation was never accepted. Nevertheless, it was considered worthy of further elaboration and was pursued by many chemists.[98]

The present situation in fundamental physics provides yet another vivid example of the difference between *acceptance* and *pursuit*. The currently accepted view on the fundamental constituents of matter and the laws that govern their behaviour is provided by the orthodox *quantum mechanics* and the *Standard Model* based on it. However, many contemporary theoreticians work on the advancement of theories which go beyond the Standard Model. Some theoreticians work on alternative quantum theories (misleadingly called "interpretations"), which should overcome the shortcomings of the orthodox theory. Others work on developing theories of quantum gravity intending to quantize general relativity. In addition, many physicists pursue one or another version of string theory (with M-theory being probably the most pursued one).[99] However, when it comes to the question of which theory is currently accepted, the answer of the scientific community is unequivocal – the orthodox quantum mechanics (with all its elaborations of course).

These examples clearly show that there is no contradiction in accepting one theory and working simultaneously in another direction. In fact, that is the only logically possible way in which science can advance. If there were no unaccepted pursued theories, nothing could possibly replace the currently accepted theories and, therefore, science would stagnate.[100]

---

[95] See Pumfrey (1989); Bennett (1989).

[96] A case can be made that a considerable chunk of the Cartesian natural philosophy was accepted during the last two decades of the seventeenth century in Cambridge, while the Newtonian theory became accepted in Oxford since the 1690s. Both of these historical hypotheses need to be confirmed, of course. The case is discussed at some length in *Part II*, section "Mosaic Split and Mosaic Merge".

[97] Among few historians who realize this, is Schmitt (1973, pp. 162–165, 179).

[98] See Laudan (1977), p. 113 and footnote 41 on p. 234.

[99] See Penrose (2004), pp. 816–1047.

[100] Nickles (2006), pp. 164–169 provides a list of 22 reasons why *pursuit* ("heuristic appraisal" in his language) should not be conflated with *acceptance* (in his language "epistemic appraisal").

To sum up, there are three different stances that the community may take towards theories and, therefore, there are three different descriptive questions:

Logic of Acceptance	Logic of Use	Logic of Pursuit
? What logic (if any) underlies the process of theory acceptance and rejection?	**?** What logic (if any) governs the assessment of the usefulness of theories?	**?** What logic (if any) governs decisions to judge a theory as worth pursuing?

It is readily seen that all three questions are formulated in purely *descriptive* terms without any *normative* ingredients. They do not ask how acceptance *ought* to take place, or how we *ought* to choose most useful or most pursuit-worthy theories. They deal exclusively with what is *actually* taking place, i.e. with the underlying mechanism of transitions from one accepted theory to another, from one useful theory to another, and from one pursuit-worthy theory to another.[101] Now, our current metatheoretical task is to clarify which of these three descriptive questions are within the scope of TSC and which are not. Should TSC account for acceptance of theories? Should TSC explain the instrumental use of theories? Finally, should TSC be concerned with scientists' decisions to pursue theories? Let us consider these meta-questions in turn.

The answer to the first question is obvious and follows from the definition of TSC as a study of changes in the mosaic of accepted theories. Any TSC should explain how transitions from one set of accepted theories to another takes place and what laws govern the process. At minimum, any TSC must explain how transitions from some accepted theories to others take place.

The answer to the second question is not as straightforward, for the definition of TSC says nothing explicit about use. However, it is sufficient for a clear answer. By definition, it is the task of TSC to explain transitions in the scientific mosaic. Again, by definition, the mosaic itself contains the theories that are accepted by the community. As for the theories that are considered useful practical tools but aren't accepted by the community, they do not constitute part of the scientific mosaic. Thus, strictly speaking, TSC is not obliged to provide a description of how a theory comes to be considered a useful tool. An actual TSC may attempt to explain the use of theories, but it may also ignore the context of use altogether.

The answer to the third question – the one regarding the logic of pursuit – can also be deduced from the definition of TSC and is similar to that of the second question. There is nothing concerning pursuit in the definition of TSC and, therefore, we are neither restrained nor obliged to provide an account of pursuit in an actual TSC.

Historically, the so-called theories of scientific method or rationality have been often involved in addressing the question of pursuit. Take Laudan's early theory,

---

[101] There are, of course, also three respective normative questions, regarding the conditions under which theories ought to be accepted (a question of methodology), used (technological question), or pursued (a socio-technological question). Evidently, none of these normative questions lies within the scope of descriptive TSC. See section "Descriptive and Normative".

for instance, in which he proposes two separate rules: one for evaluating the acceptability of a theory and one for evaluating its pursuit-worthiness. His rule of pursuit-worthiness is thus: "it is always rational to pursue any research tradition which has a higher rate of progress than its rivals".[102] Although this formulation sounds *normative* and not *descriptive*, it is in fact twofold. On the one hand, it is a rule of evaluation of pursuit-worthiness that prescribes when we ought to pursue a theory and when we ought not. Thus, it is normative. On the other hand, it is also descriptive for it supposedly describes how scientists actually decide which theories are worth pursuing.[103] Thus, we may conclude that the early Laudan assumes that it is necessary to address the descriptive question of the logic of pursuit.[104]

In his later theory, however, Laudan, seems to be ignoring the logic of pursuit altogether. In his *Science and Values*, he focuses exclusively on the logic of *acceptance* of theories, methods, and values without saying anything about *pursuit*. Another indication that the later Laudan apparently dispenses with the logic of pursuit is found in *Scrutinizing Science*. There he says that the subject of theories of scientific change is "detecting the factors that determine the acceptance and rejection of theories".[105]

By all appearances, Lakatos is among those who believe that there is no logic of pursuit. He makes it explicit that his rules are meant to cover (what I call) acceptance only. They do not apply to pursuit. Scientists, according to Lakatos, may have their own individual motives for pursuing an idea, theory, or research programme. They can even pursue a degenerating research programme if, for any reason, they find it worthy. According to Lakatos, there are no limitations in this regard.[106]

It follows from the definition of TSC, that both Lakatos's and Laudan's approaches are allowable. It is permissible to build a TSC that attempts to explain how scientists actually choose what theories to pursue, and it is equally permissible to build a TSC with no such explanation.

In summary, it is important to repeat that there is a crucial distinction between three different categories of stances that the scientific community can take towards theories. These stances are *accepted* (the opposite is *unaccepted*), *used* (*unused*) and *pursued* (*neglected*). Although these stances have often been conflated, their differentiation is a must if the positions of the scientific community are to be understood correctly.

Correspondingly, there are descriptive questions regarding each of these three stances. What is the logic of theory acceptance? What is the logic of use? What is the logic of theory pursuit? Of these three questions, as I have attempted to show, TSC should necessarily address only the first one – TSC must explain how transitions from one accepted theory to another take place and what logic governs this evolution. As for the other two questions, their discussion is not obligatory.

---

[102] Laudan (1977), p. 111.

[103] Such a conflation of normative and descriptive is very common. See section "Descriptive and Normative".

[104] See also Wykstra (1980), pp. 216, 218. Feyerabend's famous *proliferation thesis* also refers to pursuit. See Feyerabend (1975), p. 24.

[105] Donovan et al. (eds.) (1992), p. 11.

[106] See Lakatos (1971), pp. 116–117.

Individual and Social

Let us begin by clarifying the two key terms of this section. On the one hand, there is the individual scientist and her daily research. In her work, the individual scientist relies on a specific set of beliefs about the world and employs some methods to appraise the fruits of her research. Normally, beliefs and methods of the individual scientist vary with time – the individual scientist may change her views about this or that aspect of the world and may change her criteria of theory assessment. When the individual scientist changes her beliefs or methods, we have an instance of change at the *individual* level.

There is, on the other hand, the *social* level, the level of the scientific community which at any given point of time accepts a certain set of theories about the world and employs certain methods of theory assessment. Accepted theories and employed methods together comprise what we have defined as *scientific mosaic*.[107] Naturally, the mosaic also undergoes change: the community may give up some of the previously accepted elements of the mosaic and replace them with others.

Thus, we deal with two different levels of organization here:

Individual Level ≡	**Social Level ≡**
The level of the beliefs of the individual scientist about the world and the rules she employs in theory assessment.	The level of the scientific community and its mosaic of accepted theories and employed methods.

This distinction is an expression of the view commonly accepted nowadays that scientific knowledge is essentially a social phenomenon, i.e. that it only functions at the level of the scientific community. One illustration of this is the way we normally use such terms as "theory acceptance", "scientific revolution", or "scientific change". When we say "the theory is accepted", we simply mean "the theory is accepted by the scientific community". Similarly, when we speak of some transformation in science, we don't mean that this or that great scientist has changed her mind and decided to accept a new theory or employ a new method, but that the scientific community *as a whole* has rejected some elements of the mosaic and replaced them with some new elements. Finally, when we say "the method was employed at a certain period", we do not assume that the individual scientist followed that method (or that he thought he did), but merely that the method was actually employed by the community *as a whole* in theory assessment. In short, the social character of scientific change stems from the social, supra-individual character of science itself: the scientific mosaic is accepted by the scientific *community* and, thus, each and every proposed modification of it is accepted or unaccepted by the community.[108]

---

[107] See section "Descriptive and Normative".

[108] Ted Porter has expressed this same idea by pointing out that it is a peculiar characteristic of scientific knowledge to be separated from time/place and, importantly, from individuals. As he puts

Individually, each scientist has her own system of beliefs and methods which may or may not coincide with those of the scientific community of her time. Take, for instance, Galileo's heliocentrism, Einstein's famous criticism of the orthodox quantum mechanics, or Hawking's belief in the existence of God. It is obvious that in all of these cases the individual positions of these great scientists had very little in common with the content of the mosaic of their time. In more general terms, the individual scientist may believe that a certain theory is the best available theory on the market while still appreciating that it is not the one accepted by the scientific community, i.e. not part of the mosaic of the time. The individual scientist is in a position where she can legitimately say "personally, I dislike the theory in question, I don't believe a word of it, and I hope that it will be replaced by a better theory in the near future; nevertheless, I admit that it is the currently accepted theory".

Once we appreciate the distinction between the two levels, two different questions arise. At the *individual* level, the question is how exactly changes in individual beliefs and individual methods take place. What factors affect and what laws (if any) govern transitions in the views and expectations of the likes of Galileo and Einstein? At the *social* level, the question is how and why the *scientific mosaic* changes. Are there any laws that govern this process and, if so, how could they be explicated? Therefore, we have two descriptive questions:

Individual Level	Social Level
? What is the mechanism of changes in the beliefs and methods of the individual scientist?	**?** What is the mechanism of changes in the mosaic of accepted theories and employed methods?

In this metatheory, we have to determine which of these two issues should be tackled by TSC. Should TSC be concerned with changes at the social level, or should it account for transitions in individual belief systems, or both? As for the social level, the answer is straightforward: TSC, by definition, is a theory that explains transformations in the scientific mosaic. The question is, therefore, whether changes at the individual level should also be explained by TSC. Do the views of individual scientists lie within the scope of TSC, or should TSC merely focus on changes at the level of the scientific community?

Obviously, the definition of TSC is silent here: it says nothing whatsoever about the level of individual belief systems. What this means is that an actual TSC *can* but *does not necessarily have* to explain transitions at the individual level. The task of a TSC is to uncover the mechanism of transitions at the level of the scientific mosaic; if the theory also turns out to be applicable to changes in individual belief systems, so much the better, but that is not mandatory.

There is an interesting factual question that remains unanswered here: what is the actual relation between the two levels? It is clear that there must be *some* correlation

---

it, in science, one should not visit a sacred site and acquire the knowledge from the master. See Porter (1991), pp. 218–219.

between the two levels: for a change at the community level to take place, something must change at the level of individuals. This is similar to any other relation between a higher level and a lower level of organization. Just as biological processes cannot take place without the underlying physicochemical processes, social dynamics is impossible without changes at the level of individuals. So the existence of the correlation between the two levels is indisputable. What needs to be understood is the actual *mechanism* of this correlation.

Before we proceed, it must be noted that this question is beyond the scope of the metatheory and can only be properly tackled by (presumably sociological) theories of group dynamics. Thus, I won't provide any answer to this question in this metatheory. What I will try to show is that the answer to this question is far from straightforward.

Underlying many contemporary discussions is the tacit assumption that the scientific community's stance towards a theory is a straightforward function of individual stances. Individual beliefs which have nothing to do with the content of the scientific mosaic are often presented as indicative of the universal acceptance of those beliefs. Similarly, unacceptance of a theory by an individual scientist is often taken as an indication of community level unacceptance. The logic is simple: since "even Hawking believes that x", therefore, x is universally accepted. This reasoning is based on a premise that the community's stance towards a theory is a straightforward function of individual stances:

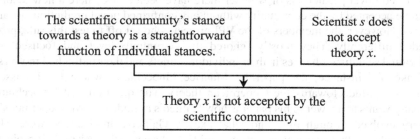

There is a similar line of reasoning concerning methods of theory assessment. Suppose we wish to find out whether method m was employed by the community at time t. How do we go about it? Some authors simply take an individual scientist, s, and inquire whether s did employ method m in her research. If it turns out that scientist s did not employ the requirements of method m in her research, then a conclusion is drawn that method m was not employed in theory assessment at time t.

Such is the strategy of Paul Feyerabend, among many others. In his attempt to show that there is no fixed and universal method of science, Feyerabend discusses several methodological dicta, such as "a new theory must conform to well-established facts" or "a new theory must be consistent with other accepted theories" and poses a question: are scientific theories indeed evaluated according to these requirements? In order to answer this question, Feyerabend discusses the case of Galileo by focusing not on the reaction of the community of the time but on that of Galileo himself. Feyerabend concludes that, in his research, Galileo clearly violated these rules. From that, Feyerabend arrives at the conclusion that the above rules have been grossly violated in the process of scientific change.[109]

Feyerabend's conclusion is clearly premature, for the question was not whether Galileo or any other individual scientist followed or violated method *m*. The question was whether method m was the one employed in theory assessment by the community. In order to answer this question, we have to determine whether, at time t, the scientific community actually proceeded in accord with method m when assessing proposed modifications to the mosaic. It is a historical fact that the implicit expectations of the scientific community of the early seventeenth century were quite different from those of Galileo. While Galileo was interested in experimental confirmations, the method employed by the community of the time required a theory to be either intuitively true or follow logically from what is intuitively true.[110] Thus, the personal opinion and practice of Galileo must not be confused with the decisions of the community of the time.

Consider another example taken from Scrutinizing Science, a collection of papers that summarizes the main outcome of the famous VPI project.[111] The task that the project set forth to accomplish was to test different methodological rules against historical episodes. One of the rules under scrutiny prescribes that, in order to become accepted, a new theory should have some confirmed novel predictions. To test whether this rule has held throughout history, the authors of the project were supposed to check whether all theories that have ever become accepted possessed confirmed novel predictions or whether there have been cases where a new theory became accepted by the community without any confirmed novel predictions whatsoever. However, the members of the VPI project, who set off to test this rule, proceeded differently. They mostly ignored the community-level and focused on individual scientists, changes in their individual beliefs and their individual motives.

Take, for instance, the paper by Maurice Finocchiaro, where he discusses Galileo's attitude towards the Copernican theory. The question that Finocchiaro actually wants to answer in his paper is how Galileo's attitude towards Copernicus's theory evolved through time and what exactly Galileo's motivations were. He clearly believes that such a clarification would be relevant to the task of their project. One conclusion that he draws is that Galileo's acceptance of Copernicanism can be traced back before the famous observations of the post-1609 period and, importantly, that Galileo's "judgement was based largely on factors other than empirical accuracy."[112] Finocchiaro emphasizes that "there is no evidence that his judgement concerning Copernicanism was based on predictive novelty."[113] This conclusion is supposed to refute the thesis that new theories become accepted only when they

---

[109] See Feyerabend (1975). Again, Feyerabend's own formulation of his conclusion is much bolder: there is no scientific method whatsoever. This conclusion, as readily seen, does not follow from his discussion.

[110] For my outline of the requirements of the late Aristotelian-medieval scientific community, see Part II, section "The Third Law: Method Employment", pp. 139 ff.

[111] Donovan et al. (eds.) (1992). For discussion of the project, see Nickles (1986, 1989); Richardson (1992).

[112] Finocchiaro (1992), p. 56.

[113] Ibid.

provide confirmed novel predictions. Similar lines of reasoning are present in many other papers of the volume.

It is apparent, however, that this argument is flawed. Instead of discussing the acceptance/unacceptance of the Copernican theory by the scientific community, the author focuses on one scientist, Galileo. He shows (quite convincingly to be sure) that the behaviour of Galileo is at odds with the rule under scrutiny. However, the conclusion that is drawn from the discussion is different: in the Summary of Results towards the end of the volume, Finocchiaro's analysis is taken as disproving the rule which says that a new theory must have confirmed novel predictions in order to become accepted. The rule, therefore, is said to be refuted by the historical record.[114]

Evidently, both Feyerabend's and Finocchiaro's arguments exemplify the same pattern. Instead of showing that the rule in question was not employed in theory assessment by the community, they attempt to show that the rule was not employed by individual scientists. Their reasoning is based on a tacit conviction shared by many contemporary authors that the community's acceptance/unacceptance of a proposed modification to the mosaic is a straightforward function of individual acceptances/ unacceptances. Changes in the scientific mosaic, in this view, are a function of changes in the belief systems of individual scientists. As Nickles puts it, "according to this conception, community decisions are a function only of individual decisions, which are more or less punctiform, datable historical events and are entirely the result of deliberation about information to which the individual has conscious, subjective access – rather than something that works itself out in historical discussion and other activity at the community and higher levels."[115] It is this premise that most of the members of the VPI project, as well as many other authors, tacitly subscribe to.[116]

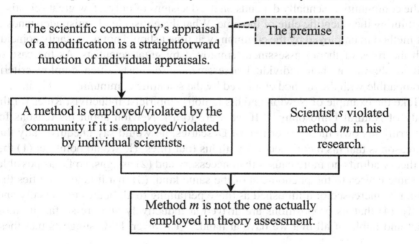

---

[114] Donovan et al. (eds.) (1992), pp. 381, 377.

[115] See Nickles (1989), p. 668.

[116] Kuhn shares this conviction. One question he is concerned with is why Kepler and Galileo were early converts to Copernicus's system. Kuhn seems to think that by answering this question he would explain the community's reaction to Copernicus's theory. See Kuhn (1977), pp. 324–329, 332, 334.

This fundamental premise may be read in at least two ways. On the one hand, it may be taken as saying that the method of science (if there is any) is exemplified in the actions of great scientists. In this reading, it is great individuals whose decisions stand as exemplars of scientific rationality. Where else should we look for instances of genuine scientific reasoning, it is argued, if not in the decisions of the likes of Galileo and Einstein? This elitist reading is championed, among others, by Lakatos who suggests that it is the value judgements of the scientific elite that should be taken as instantiations of the scientific rationality and, thus, the methodological rules are to be tested against the judgements of the elite.[117]

On the other hand, the premise may be understood as saying that the community's acceptance of a proposed modification in the mosaic is merely a sum of the acceptances of all individual scientists working in the field. In this view, a proposed modification is being accepted when a majority (either simple or super-) of the members of the community opt for it. Correspondingly, a method is said to be employed by the community in theory assessment when it is employed by a majority of the scientists working in the field. This reading can be called majoritarianist. Laudan, together with most of the authors of the VPI project, seems to be among the proponents of this view.[118]

However, we have good reason to believe that acceptances/rejections at the level of scientific mosaic are not a straightforward function of individual acceptances/rejections. There must be, of course, a certain correlation between individual beliefs/methods and the theories/methods of the scientific mosaic, but apparently this correlation is more subtle than both elitists and majoritarianists allow.

Consider first the elitist view. The elitist submits that the employment of a method by the community essentially depends on the decisions of only a few great scientists, constituting the scientific elite of the time. What the elitist clearly misunderstands is that method m employed by the community at time t should not necessarily coincide with the rules of theory assessment employed by great scientists. As the case of Galileo shows, in their individual assessments, scientists often employ criteria incompatible with the method employed by the scientific community of the time.

Take the example of Newton and his famous empiricist-inductivist Rules for the Study of Natural Philosophy.[119] If we were to ask Newton about his reasons for preferring his own theory over that of Descartes, he would say unequivocally that the reason is that his theory accords with his four rules. He would point out (1) that his theory admits no more causes than necessary and (2) assigns, so far as possible, the same causes to the phenomena of the same kind, (3) that it takes qualities that cannot be increased or diminished as those pertaining to all bodies universally and, finally, (4) that its propositions are arrived at inductively and are so far the most exact and liable of all that we have at hand.[120] However, HSC suggests that these

---

[117] See Lakatos (1971).

[118] See Laudan et al. (1986), p. 160. See also Laudan's account of the scientific method of the eighteenth and early nineteenth centuries in Laudan (1984), pp. 55–60.

[119] See Newton (1687), pp. 794–796.

[120] For Newton's methodology, see Smith (1989).

were not the rules employed by the scientific community at the time when the Newtonian theory finally became accepted. When we study carefully the circumstance of the acceptance of Newton's theory, it becomes apparent that it was not until after its prediction of the oblate-spheroid shape of the Earth became confirmed by the Lapland and Peru expedition that the Cartesian physics was rejected and the Newtonian theory became accepted by the community as a whole (and not only by British scientists). The proponents of both the Cartesian and Newtonian theories saw it necessary to test the predictions of the rival theories by observations. What is important here is that the Newtonian theory was declared advantageous not by Newton's own standards, but by the standards of a somewhat different hypothetico-deductive method, i.e. Newton's theory became accepted not because it was "arrived at inductively" (whatever that can possibly mean), but simply because one of its novel predictions was confirmed.[121] Obviously, this method is in sharp contrast with Newton's own four rules. The conclusion that we should draw from this example is that great scientists do not necessarily subscribe to the method actually employed by the community in theory assessment.

Or take a more recent example – Roger Penrose and his version of twistor theory. Undoubtedly one of the leading specialists in his field, Penrose adheres to his version of Twistor Theory, one of several existing attempts of uniting general relativity and quantum mechanics. We do not need to go into details on how exactly the theory marries up quantum mechanics and general relativity.[122] What is important for our purpose is that, albeit not without support, Penrose's theory is not the one accepted by the community. Again, we have an example of a great scientist deviating in his views from those of the community. This discrepancy between individual beliefs/ methods and the scientific mosaic is a very common historical phenomenon.

Thus, we have to conclude that the elitist view is way off the mark: individual beliefs and methods of the scientific elite are not always indicative of the state of the mosaic of the time. It is conceivable that a theory or a method may be part of the mosaic even when the members of the elite individually subscribe to other theories or employ other methods. In other words, there is more to the community's acceptance than the elitist allows.

Let us now turn to the majoritarianist view. On the majoritarianist account, a theory is accepted and a method is employed by the community when the majority of scientists involved in the field accept/employ it. It actually provides quite a straightforward picture of the correlation between the individual and social levels – the decision of the community is nothing but the decision of a majority.

Again, we have good reason to suspect that this majoritarianist picture is not quite correct. Contrary to the majoritarianist view, it has often happened that a theory was accepted by the scientific community even when most of the scientists involved in the field personally preferred other theories. Suffice it to consider two

---

[121] It's no wonder that it was Maupertuis, arguably the most vivid advocate of the Newtonian theory in France (Voltaire aside), who insisted on the necessity of testing the predictions of the Newtonian theory concerning the shape of the Earth. For details see Terrall (2002).

[122] For details see Penrose (1999). For a popular account see Penrose (2004), pp. 958–1009.

famous historical cases – the status of the Aristotelian natural philosophy in the second half of the seventeenth century and the contemporary status of the orthodox quantum mechanics.

It is a well-known historical fact that, in France, the Aristotelian-medieval natural philosophy was eventually replaced in the scientific mosaic by the Cartesian mechanical natural philosophy only circa 1700. Yet, it is safe to say that as early as the mid-1670s a majority of the scientists in the Académie openly accepted one or another version of the mechanical natural philosophy, mostly that of Descartes and his followers.[123] It can be cautiously suggested that in the second half of the seventeenth century the Aristotelian-medieval natural philosophy didn't have the support of the majority of individual scientists. However, the record shows that the Aristotelian-medieval natural philosophy continued to be taught in French universities until the end of that century. The Cartesian natural philosophy was even condemned by the leading universities despite all the attempts of its proponents to squeeze it into the curriculum. It was not until the late seventeenth and early eighteenth centuries that the Cartesian natural philosophy began to be taught in some universities.[124] This suggests that a theory can remain accepted even when it does not have the support of the majority.

One may rightly point out that my reading of the history of this period may well turn out to be mistaken for one simple reason – the statistics on the opinions of individual scientists of the seventeenth and eighteenth centuries is not easily obtainable. Even when we deal with the most studied scientists, such as Newton, Boyle, or Huygens, we are not always in a position to indicate their respective views regarding this or that issue with the necessary precision. Let us therefore turn to a contemporary case, where the statistics of the individual opinions is readily obtainable.

The standard picture that is portrayed in the contemporary literature on the status of quantum mechanics and its so-called "interpretations" is this. The theory which is accepted nowadays is the so-called "Copenhagen interpretation" and, it is said, there are several competing "interpretations" which strive to replace the orthodox "Copenhagen interpretation".[125] Now, let us first clarify that the so-called "interpretations", such as Copenhagen, many worlds, Bohm, consistent histories etc., are not interpretations at all, but in fact full-fledged theories.[126] Therefore, the real picture is that the accepted theory – we can call it orthodox quantum theory – is being challenged presently by several competitor-theories, such as the many worlds theory, Bohm theory, consistent histories theory etc. What is interesting from our standpoint is that the orthodox quantum theory remains accepted despite the fact that

---

[123] See Armitage (1950); McClaughlin (1979).

[124] See McClaughlin (1979), p. 569; Brockliss (2003), pp. 45–48.

[125] See, for instance, Albert (1992); Weinberg (1992); Ghirardi (2005).

[126] They are called "interpretations" as a kind of tribute to logical positivism and its views on theory and interpretation – a view, rejected long ago. In particular, the view assumed that theories are mere mathematical formalisms; interpretations are something different. We do not have to repeat this clumsy mistake for the sake of preserving the long-abandoned tradition.

only a minority of the scientists involved in the field actually believes in the theory. Theoretical physicists have been polled as to their preferred quantum theory on several occasions. Although the polls are usually informal and give varying results, one important parameter remains stable from poll to poll: only a minority of the theorists believe that the orthodox theory is the best we have. A majority of the polled, including such scientists as Stephen Hawking and Murray Gell-Mann, opt for one or another of the competitors (the most favoured usually being the many worlds theory).[127] At the same time, nobody seems to deny that the orthodox quantum theory is still the accepted one. An indication of this is the fact that students of theoretical physics throughout the world are taught that theory as the best available on the market. This outcome is extremely troubling for the majoritarianist; it seems to suggest that the acceptance by the community is not a straightforward function of the individual acceptances. Thus, the correlation between community's acceptance/unacceptance and decisions of individual scientists is not as simple as the majoritarianist assumes.

As for the question of the actual mechanism of this correlation, it is not to be tackled by this metatheory. Here I can only add that there need not be anything mystical about that correlation: there is no need to resort to Hegelian or Durckheimian conceptions of methodological holism as Nickles appears to suggest.[128] It is possible that the acceptance of a theory at the social level is a function of individual beliefs about what is accepted, i.e. it is possible that a theory is accepted not when a majority of scientists think it is the best description of its object, but when they think the theory is commonly accepted. Alternatively, it may turn out that two different individual-level mechanisms are responsible for the process of becoming accepted and the process of maintaining the acceptance. It might be the case that theories become accepted only when they have the support of a majority (or supermajority) of individual scientists, while they often remain accepted even without the support of a majority. This would be analogous to the situation in politics: the support of a majority of voters is needed to win the seat, but is not always required to remain in the seat.[129] Of course, it is equally possible that the actual correlation between the two levels is much more subtle than this. In any case, I would like to emphasize that I am not suggesting any particular solution here, for only an actual study of that correlation between the two levels can reveal its mechanism.

What is crucial from the metatheoretical perspective is that TSC is to study changes in the mosaic accepted by the community and not to confuse them with individual motives and decisions. To that end, we have to focus on those sources which are indicative of the mosaic itself. Luckily, there are such sources: textbooks,

---

[127] See Price (1995); Tegmark (1998).

[128] In this I agree with Brown. See Brown (2001), p. 141; Nickles (1989), p. 668. An alternative mechanism – the so-called "joint acceptance" model – is presented in Gilbert (1987).

[129] The idea that the support of a majority might be required only when a theory becomes accepted but not necessarily while accepted was suggested by Daniel Carens-Nedelsky during the seminar of 2013.

university curricula, encyclopaedias, to name a few.[130] In any case, if we wish to find out what theory was accepted at a specific point of time, personal letters and individual confessions are probably not the best place to look for an answer. The two levels must not be confused.

Luckily, many authors realize the difference between the two levels. This difference has been repeatedly emphasized by Miriam Solomon.[131] Robert Nola and Howard Sankey also focus on the acceptance by the scientific community when they ask "what determines the historical sequence of scientific theories that were actually chosen by the community at large?"[132] The distinction between the individual and social levels has also been drawn by Ernan McMullin. That the acceptance by the scientific community is not a sum of the preferences of individual scientists is also implicit in Stephen Brush's discussion of the acceptance of Mendeleev's theory.[133]

Let us now sum up the key points of this section. It is implicit in the definition of TSC that it should explain changes in the scientific mosaic of accepted theories and employed methods, i.e. changes at the level of the scientific community. As for the individual level, TSC is not charged with the immediate task of explaining changes in individual beliefs/methods. Of course, an actual TSC may turn out to be also applicable to the individual level, but this is not required from a TSC. Albeit an important issue in itself, the study of the evolution of individual belief systems is not the task of TSC. The reason for distinguishing between the two levels is that the correlation between them is all but simple. On the one hand, great scientists do not necessarily accept those theories which are accepted by the community. On the other hand, it is possible for a theory to be accepted by the community with most of the scientists personally accepting some other theory. Thus, the study of individual opinions and decisions should not substitute for the real task of TSC – the study of changes in the scientific mosaic.

Explicit and Implicit

Having discussed the difference between the individual and social levels, we shall now move on to another important distinction which unfortunately has been often overlooked. I have been repeating that the scientific mosaic consists of accepted theories and employed methods. But what exactly do we mean by "method"? In the context of theory appraisal, "method" has been traditionally assigned two different meanings. On the one hand, "method" has been taken as a set of rules of theory

---

[130] For discussion of indicators of theory acceptance and method employment see section "Indicators" below.

[131] Solomon (1992), p. 452 and Solomon (1994).

[132] Nola and Sankey (2000), p. 7.

[133] See McMullin (1988), p. 23 and Brush (1994), p. 140.
    Unfortunately, not all authors draw this distinction. See, for instance, Wykstra (1980), pp. 213, 215; Lugg (1984) p. 436; Knowles (2002), p. 177.

appraisal explicitly professed by scientists. On the other hand, "method" has also been understood as a set of rules implicitly employed in actual theory appraisal. Unluckily, these two different meanings have been often confused. Yet, it is easy to see that the rules actually employed in theory assessment and the rules openly prescribed by the community are essentially two different types of entities.

Consider a scientific community which openly prescribes certain rules for theory assessment, say, they prescribe that in order to become accepted a new theory ought to solve more problems than its predecessor. Suppose, also that when it comes to actual theory assessment this community accepts only those new theories which have confirmed novel predictions regardless of the number of solved problems. From this, we can conclude that the open prescriptions of this community have little to do with their actual expectations, their actual practice of theory assessment. There is a significant difference between their explicit and implicit rules. This shows us that the actual implicit expectations of the community may or may not coincide with their own open prescriptions.

Once we appreciate that the two do not necessarily coincide, we must also appreciate that the two categories need two different labels. Henceforth, "method" will be short for "implicit rules of theory assessment", while "methodology" will be short for "explicitly stated prescriptions":

Method ≡	Methodology ≡
A set of implicit rules to be employed in theory assessment.	A set of explicitly formulated rules of theory assessment.

Clearly, it is one thing to say that a certain set of rules was openly prescribed by the scientific community and it is quite another thing to insist that this set of rules was the one employed by that community in actual cases of theory assessment. The rules of the methodology openly stipulated by the scientific community may or may not be the same as the implicit rules of the actual method employed by the community. Indeed, the rules of the actually employed method may differ drastically from the methodological dicta explicitly stated in textbooks and encyclopaedias. This should not come as a surprise. In fact, as Steven Weinberg has recently pointed out, "most scientists have very little idea of what scientific method is, just as most bicyclists have very little idea of how bicycles stay erect."[134]

Consider some historical examples. It is well known that during the second half of the eighteenth century and the first half of the nineteenth century the scientific community explicitly prescribed a version of the empiricist-inductivist methodology (championed by Newton no less), which stipulated that new theories should be "deduced from phenomena" and that they should not postulate any unobservable ("occult") entities or qualities. Theories, that contained theoretical terms referring

---

[134] Weinberg (2003), p. 85.

to unobservable entities (such as atoms or invisible fluids), were officially considered unacceptable. It is safe to say that this empiricist-inductivist methodology was openly prescribed by the scientific community of the eighteenth century.[135] However, the historical record also shows that during the era of the dominance of this methodology, several theories that postulated unobservable entities somehow managed to become accepted by the scientific community. For one, there was Franklin's theory of electricity which postulated the existence of unobservable electrical fluid responsible for many electrical phenomena. There was also the then-accepted chemical theory which explained the processes of combustion and rusting of metals by postulating that all flammable materials contain phlogiston, an unobservable substance without odor, taste, color, or mass. Yet another example is Fresnel's wave theory of light that postulated the existence of luminiferous ether, a medium for the propagation of light – a concept which could by no means be "deduced from phenomena". The most striking example, however, was Newton's theory itself, for it postulated the existence of such unobservables as gravitational attraction, absolute space, and absolute time. What all these cases clearly demonstrate is that the actual expectations (the method) of the scientific community of the time had little in common with the rules of the openly prescribed empiricist-inductivist methodology. History gives us a great many examples of this sort.

At this point, a terminological clarification is in order. To avoid possible confusion, I suggest that we reserve the adjective "accepted" exclusively for theories. In this sense, we can speak of the currently accepted theories in physics, chemistry, biology, sociology etc. Similarly, we can speak of the accepted natural philosophy circa 1650 or the accepted theology circa 1550. As for methods, I suggest we use adjective "employed": methods can be said to be employed in theory assessment (but not accepted). Thus, we cannot speak of "accepted methods", for only theories can be accepted. Methods, i.e. implicit expectations of the community, can only be employed in theory assessment. It is my suggestion to keep the terminology as clear as possible in order to avoid conflations abundant in the contemporary literature on scientific change. Here is the definition of employed method:

Employed Method ≡
A method is said to be *employed* at time t if, at time t, theories become accepted only when their acceptance is permitted by the method.

The opposite of employed is not employed, or unemployed. Obviously, a method which is actually employed in theory assessment nowadays need not be the exact same method employed in theory assessment 300 hundred years ago. Similarly, the implicit expectations of the Aristotelian-medieval scientific community need not coincide with the implicit expectations of, say, the scientific com-

---

[135] There are many indications of this. See d'Alembert (1751).

munity of Paris circa 1740. Moreover, it is a historical fact that they did not coincide. The fact that the Newtonian theory became accepted by the scientific community of Paris only after the confirmation of one of the theory's novel predictions (circa 1740) is a good indication that the requirement of confirmed novel predictions was part of the method employed by the community of the time – it was one of the implicit expectations of the scientific community of Paris circa 1740. This requirement, however, wasn't always part of the scientific method. If we went back to the 1500s, we would notice that theories in natural philosophy were not expected to have any confirmed novel predictions: the Aristotelian-medieval scientific community had different implicit expectations, i.e. a different method. As a first approximation, it might be argued that the Aristotelian-medieval scientific community would only accept a theory if it appeared to be grasping the nature of a thing through intuition schooled by experience, or if it was deduced from general intuitive propositions.[136] In short, there are methods that have been employed but are no longer employed in theory assessment. Thus, at any moment of time we have a picture similar to this:

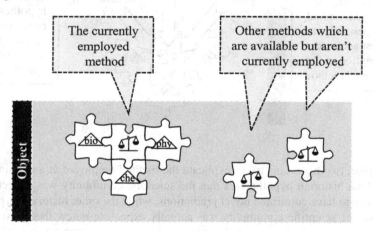

The historian that studies a certain time period needs to keep in mind the difference between method and methodology. Describing the official methodology of the community of the time is clearly not sufficient, for the primary task is to dig deeper and try to unearth the actual expectations of that community, i.e. their employed method.[137] Importantly, when the historian attempts to unearth the rules of the employed method of the time, she thus proposes a historical hypothesis which may

---

[136] I provide my own explication of the Aristotelian-medieval method in Part II, section "The Third Law: Method Employment".

[137] There is an important question of how the implicit expectation of a given time can be uncovered. I will address this question in section "Indicators" below.

or may not be correct. For instance, if I claimed that, nowadays, the employed method of physics requires a theory to have confirmed novel predictions in order to become accepted, I would be proposing a historical hypothesis which purportedly describes the actual expectations of the community. Naturally, as with any other hypothesis, my description of the employed method could become accepted by the community and become part of our historical knowledge, or it could be considered incorrect by the historians and remain unaccepted. Other historians may propose their own explications of the currently employed method. If eventually one of the available historical hypotheses concerning the contemporary method of physics becomes accepted by the community, we will have the following situation:

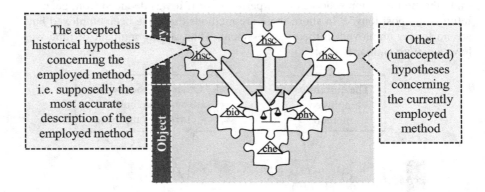

Suppose, two historians try to explicate the method employed at a certain time period. One historian hypothesizes that the scientific community was expecting a new theory to have confirmed novel predictions, while the other historian hypothesizes that the scientific community was actually expecting a new theory to solve more problems than the accepted theories. Now, suppose that neither of these hypotheses is correct, for in reality the community expected common sense explanations of known facts. Yet, it is conceivable that one of these hypotheses becomes eventually accepted by the community. If, say, the first hypotheses became accepted by the community, we would have the following picture[138]:

---

[138] In order to distinguish between methods and theories in our diagrams, I suggest we represent methods in dashed rectangles and theories in solid rectangles. Dashed rectangles indicate that methods are often not on the surface.

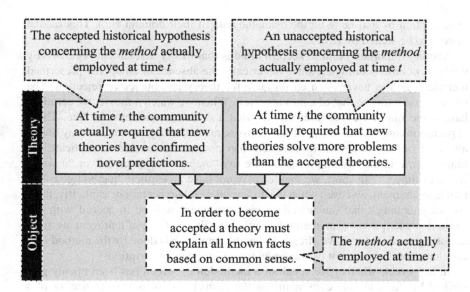

The situation with methodologies is similar. When studying a certain time period, the historian may come across many different methodologies that were available on the market. Sometimes, the historian may also find that one of these methodologies was commonly considered as the right way of doing science. Recall, for example, the rules of the Aristotelian-medieval methodology before the Scientific Revolution, or the empiricist-inductivist methodology in the eighteenth century, or the logical positivist methodology in the first half of the twentieth century. Albeit at different periods, each of these methodologies has been commonly considered as prescribing the right way of doing science. In contrast, there have been many other proposed methodologies that have never been officially prescribed by the community. The twentieth century alone produced so great a number of methodologies that even the most thorough of our textbooks cannot mention all of them.[139] Alternatively, our historian may find out that, at the time period under study, there were many competing methodologies but none of these methodologies was officially prescribed. It might be tempting to argue that we are witnessing such a situation presently in the physics community. Our textbooks, as some contemporary analyses show, often differ substantially in their methodological dicta.[140] However, it may be that, despite all discrepancies, there are nevertheless some basic (if somewhat vague) methodological principles openly prescribed by the community (e.g. that physical theories ought to be tested in repeatable experiments and observations, or that physical theories

---

[139] See, for instance, Nola and Sankey (2007).
[140] See Blachowicz (2009).

must explain by and large all the available data of their domain etc.). This factual issue is to be settled by HSC.

Another important point to keep in mind is that theory assessment is carried by methods not methodologies, since historically the absence of an openly prescribed methodology has never been an obstacle for theory acceptance or rejection. The situation is similar to that of a movie review. When we watch a movie, we certainly have some implicit expectations as to what a decent movie must be like. These expectations are our "method" of movie assessment. Clearly, we may or may not be aware of our expectations. More often than not, we are unable to explicate our implicit criteria that we employ when we say "the movie was great!" or "it was a waste of time...". In short, we may or may not have an explicit "methodology" of movie assessment, and even when we do formulate our expectations explicitly, there are no guarantees that our actual movie assessment will be in accord with our explicit prescriptions (our methodology). Luckily, that doesn't prevent us from assessing movies, for as we already know the actual job is done by the method (our implicit expectations), not by methodologies (our open prescriptions).

In any event, whether we speak of a methodology which has been openly prescribed by the scientific community as the correct way of doing science, or of a methodology which has never gained any popularity, we should remember that we speak of a set of openly formulated rules which are not necessarily the same as the rules actually employed by the community in theory assessment. Method and methodology should not be confused.

The idea that the two concepts – method and methodology – should be distinguished is not new. It can be traced back to Albert Einstein who advised that in order to explicate the method of physics one should attend to what physicists do (method, in my terminology), and not to what they say they should be doing (methodology).[141] Similarly, both Carnap and Popper realized that what is employed in theory assessment is the actual method, rather than the open prescriptions of the community. The distinction between the two is also implicit in Lakatos's conception. According to Lakatos, methodologies (such as inductivism, conventionalism, or falsificationism in all their different variations) may or may not coincide with the implicit method of science. The distinction is vital also for the later Laudan's reticulated model, where one way of criticizing the accepted views on scientific practice is by showing that the accepted rules (methodology) are at odds with the actual scientific practice of the time (method).[142]

---

[141] See Einstein (1934). This dictum has been restated on many occasions. See, for instance, Westfall (1971), p. 41; Wykstra (1980), p. 211; Lugg (1984), pp. 436–438.

Whether one indeed should refer to "what scientists do" in order to uncover the actual method is a different issue. See section "Indicators" below.

[142] See Laudan (1984), pp. 56–61, 82–84. Among those who emphasize the difference are also Wykstra, Worrall, and Lindberg. See Wykstra (1980), pp. 211–212; Worrall (1988), p. 266; Lindberg (2008), p. 362.

However, Laudan's position is somewhat confusing. On the one hand, he seems to distinguish between method and methodology on several occasions.[143] On the other hand, when it comes to explaining the actual mechanism of scientific change, Laudan forgets about changes in methods and only focuses on changes in methodologies. He devises strategies for critical evaluation of explicit methodological rules, but he says very little about actual transitions from one employed method to another.[144] Even those who clearly see the difference between the two concepts often use "method" and "methodology" interchangeably. Zahar, for instance, uses "presystematic methodology" when he speaks about requirements actually employed by scientists.[145] Even such a cautious analyst as Lakatos sometimes substitutes "implicit methodology" for "implicit method".[146] Yet, many authors do not seem to distinguish between the two concepts at all. Leplin, for instance, seems to be ignoring the distinction when he says that Newton's methodology was actually employed in theory assessment during the 1810–1820s.[147] In any case, it is obvious that the two should not be mixed up.

If we combine this distinction with that between individual and social, we will obtain four different categories:

	Method	Methodology
Social	The rules *actually* employed by the *scientific community* in theory appraisal.	The rules of theory appraisal *openly* prescribed by the *scientific community.*
Individual	The rules *actually* employed by the *individual scientist* in theory appraisal.	The rules of theory appraisal *openly* prescribed by the *individual scientist.*

The metatheoretical question that must be answered here is transitions in which of the above four must be traced and explained by TSC. The discussion in the previous section reveals the answer concerning the individual level: an actual TSC may

---

[143] See, for example, Laudan (1968), p. 4.

[144] Lugg has pointed this out in his review of Laudan (1981). See Lugg (1984), pp. 436–437.

[145] See Zahar (1982), p. 407.

[146] See Lakatos (1971), p. 120; Worrall (1988), p. 266.

[147] See Leplin (1997), p. 38. Iliffe sounds along the same lines in his (2003), p. 272. Brush too doesn't distinguish the two in his (1994), p. 140. See also, Newton-Smith (1981). The confusion can be traced back to Whewell (1860).

successfully explain changes at the individual level but it isn't required to do so. Thus there are only two questions concerning the scope of TSC that remain to be tackled here. Firstly, should TSC account for changes in methods employed in theory assessment? Secondly, should TSC explain transitions from one openly prescribed methodology to another? I shall argue that the answer to the first question is affirmative. As for the latter question, the answer is currently indeterminate and will depend on how the problem of the status of normative propositions is solved.

While changes in methods are clearly within the scope of TSC, the status of methodologies in the mosaic is currently unclear. This has to do with a more general issue of the status of normative propositions. At the moment, it is uncertain whether normative propositions (value judgements, prescriptions etc.) can be part of the scientific mosaic. It is obvious that commonly prescribed normative propositions can be found at any time period. Nowadays, for instance, it is commonly agreed that we ought to be tolerant towards different cultures insofar as these cultures do not themselves promote cultural intolerance. It is however unclear whether such normative propositions are part of the scientific mosaic. The same uncertainty applies to methodological dicta. We clearly share some methodological principles, but it is unclear whether they actually affect the process of scientific change. This question cannot be settled at the level of metatheory; only an actual TSC together with HSC can tell us whether normative propositions in general and methodological propositions in particular can be part of the scientific mosaic. At this time, we can only say that at minimum any TSC must explain changes from one employed method to the next.

It is worth mentioning that, while there have been many attempts to explain transitions from one physical theory to the next or from one biological theory to the next, so far there have been very few serious attempts to account for transitions in methodologies and even fewer attempts to explain changes in actual methods. Lakatos provided a theory that was supposed to explain transitions from one methodology to another.[148] But he didn't say anything about changes in methods for, as almost all authors before him, he too considered the method of science unchangeable.[149] Kuhn's conception of paradigms and scientific revolutions too can be taken as providing an explanation of changes in methodologies.[150] It could also be understood as applying to changes in methods as well, although, as is often with Kuhn, his position on the subject is vague. Finally, the later Laudan's reticulated model attempts to cover changes in methodologies.

The case of the later Laudan, however, is the most puzzling and should be considered separately. On the one hand, Laudan clearly understands that the actual practice of theory assessment (method, in our terminology) and explicitly stated methodological dicta are not the same. In fact, it is a paramount idea of the later Laudan that we gradually modify our views on scientific practice, in one way by

---

[148] See Lakatos (1971).

[149] But see note 22 above.

[150] See Kuhn (1962/70).

identifying apparent discrepancies between the actual scientific practice and our explicit views on that practice.[151] On the other hand, the later Laudan's theory does not say anything about transitions in actual scientific practice (i.e. employed methods), and focuses exclusively on transitions from one openly prescribed methodology to the next. The situation is quite puzzling, for he couldn't fail to see that it isn't the methodology but the actual method that does the job – theories are assessed by the actual method, not by what is officially proclaimed as the correct rules of assessing theories. Had Laudan held that openly prescribed methodologies and actually employed methods are essentially two sides of the same thing, his neglect of changes in methods would have been natural, for in that case he could have claimed that the same set of laws governs changes in both methodologies and methods. But he clearly realizes that our explicit proclamations (methodologies) and our implicit requirements (methods) are not the same. I confess that I do not see how Laudan could both admit that method and methodology are not the same and nevertheless neglect changes in methods.

There are two conclusions to be drawn from the discussion of this section. First, the rules of the method actually employed in theory assessment may or may not coincide with the rules of the openly prescribed methodology. Method and methodology are two different entities and are not to be conflated. Second, transitions in methods are to be accounted for by TSC for it is actual methods that are employed in theory appraisal. As for changes in methodologies, at this stage it is unclear whether they are within the scope of TSC. The status of methodologies is uncertain, as is the status of all normative propositions in general; if normative propositions turn out to be part of the mosaic, then transitions in methodologies will have to be explained too.

Time, Fields, and Scale

So far we have clarified that TSC is a descriptive theory concerned with transitions from one accepted theory to another and from one employed method to another. This section concludes our clarifications of the scope of TSC. There are three important questions that need to be answered here. First, for changes in the mosaic of what time period should TSC account? Secondly, for changes in which fields of inquiry should TSC account? Finally, should TSC explain only grand changes, or should it also account for minor changes? Let us consider these questions in turn.

---

[151] Laudan's favourite example is the transition from the then-accepted empiricist-inductivist methodology to the methodology of hypothetico-deductivism. The transition, according to Laudan, took place in the first half of the nineteenth century and was due to the recognition of the fact that the actual practice of science was at odds with the empiricist-inductivist methodology; namely the fact that many theories postulating unobservable entities became accepted during the era of dominance of the empiricist-inductivist methodology. See Laudan (1984), pp. 55–59.

The first question refers to the historical time period that TSC should deal with. Should TSC explain all transitions that the mosaic has undergone since antiquity, or should it only cover changes starting from some later period (e.g. the seventeenth century, the early twentieth century)?

Some of the most prominent conceptions of scientific change can only be understood as attempts to account for transitions in the post-sixteenth (or even post-seventeenth) century science. Take, for instance, Lakatos's three rules of theory appraisal which stipulate that a modification within a research programme is progressive if (1) it explains the success of a previous theory, (2) predicts novel hitherto unexpected facts and (3) has some of its novel predictions confirmed.[152] Obviously, these rules can hardly be taken as a correct description of the actual practice of, say, ancient or medieval science. Indeed, one can hardly be serious in arguing that, say, in 400 BC or 1400 CE novel predictions played the same role as in 2000 CE. Lakatos's rules can at best be taken as an attempt to explicate the actual method of science employed after the Scientific Revolution. For instance, it is obvious that the requirement of confirmed novel predictions wasn't part of the Aristotelian-medieval method.

The same can be said with regard to the position of the VPI project. It is assumed by the members of the VPI project that the general theses under scrutiny refer only to the post-sixteenth century science. Suffice it to look at the list of the examined historical episodes: none of the episodes concerns changes in ancient or medieval science. In addition, there is direct textual evidence. In their programmatic paper, the members of the project openly state that their focus is the post-sixteenth century science.[153]

This lack of attention to the pre-sixteenth century science is understandable if we take into account that prior to the 1980s relatively little attention was paid to the history of ancient and especially medieval science, a field of inquiry that has witnessed astounding growth during the last three decades. This fact, however, did not stop authors like Kuhn from proposing models of scientific change which supposedly applied to all periods of history. Kuhn's model of transitions from a pre-paradigm period to the successive periods of normal science and scientific revolutions is one such example.[154] It was meant to explain transitions not only in modern, but also in ancient and medieval science.

Laudan's both early and late theories can also be viewed as attempts of accounting for all historical periods.[155] Take his early theory which stated that competing

---

[152] See Lakatos (1970), pp. 32–34.

[153] See Laudan et al. (1986), p. 149. It has been even suggested that one need not go that far and may legitimately confine oneself to studying relatively recent scientific practice. Ronald Giere, for example, seems to be saying that in order to understand the workings of science one need not delve deep into its history; it would be sufficient to study the contemporary scientific practice. See Giere (1973), p. 290 and Giere (1984), p. 28. In fact, his "model of scientific development" with its explicit reference to experimentation can hardly be applicable to the Aristotelian-medieval period. See Giere (1984), p. 25.

[154] It is important to take Kuhn's (1962/70) together with its (1977) patch.

[155] Laudan's early theory is presented in his (1977), late theory in his (1984).

theories are compared by their overall problem-solving capacity. According to this theory, theories become accepted when they solve more problems than their competitors. It is readily seen that this model is applicable to the Aristotelian and Newtonian science alike (chiefly because it is not strict enough nor does it say much; but this is a different issue). The same goes for Laudan's later theory. His reticulated model of scientific change is readily applicable to both the pre- and post-sixteenth century science. Certainly, it fails to correctly describe the actual mechanism of scientific change, but the point is that it attempts to do so, i.e. it does not limit its applicability only to post-sixteenth century science.

Now, the question that has to be addressed here is whether we expect TSC to cover all historical periods or whether it should confine itself to the time period since the seventeenth, eighteenth, or twentieth centuries. In other words, does the scope of TSC have some temporal constraints? If we refer to the definitions of scientific change and TSC, the answer becomes obvious: TSC should account for all changes in the scientific mosaic, since the definition says nothing about temporal limitations.

Of course, in practice, the situation is more complex: while at some time periods there seem to be more than one mosaic, at other periods even a single mosaic is hard to locate. In some cases, it appears that we cannot speak of the scientific community but at best only of different, very loosely related scientific communities and, thus, different scientific mosaics. This is especially true of the ancient and early medieval periods. Even as late as the eighteenth century, there existed at least two different scientific communities – one on the Continent and one in Britain (roughly) – each with its peculiar scientific mosaic. To skip the details, both communities had rejected the Aristotelian natural philosophy at approximately the same time, circa 1700. However, while in the British scientific mosaic the Aristotelian natural philosophy was replaced with that of Newton, in the Continental scientific mosaic it was replaced by the Cartesian natural philosophy.[156] In other cases, it is not clear whether we can properly speak of a scientific community at all. For instance, when we consider the tenth century Europe, it is hard to tell whether there was any scientific community and therefore any mosaic of accepted theories and employed methods to speak of. For the purposes of TSC, it would suffice if there were at least some accepted theories and some employed methods. But were there any? This is an important factual question which is to be addressed not by TSC, but by HSC. If, for instance, HSC were to suggest that at some time period there did exist a certain scientific mosaic, then we would be in a position to speak of the scientific community of that period. In that case, TSC would have to account for changes in that scientific mosaic. If, however, it turns out that we cannot properly speak of the sci-

---

[156] The indications of their dissociation are very well known and are included in standard narratives of the Scientific Revolution. Perhaps the most prominent piece is Voltaire's Letters on England and especially often quoted Letter XIV entitled "On Descartes and Sir Isaac Newton", where Voltaire depicts the key differences in the views of the two scientific communities. See Voltaire (1733), pp. 68–72.

In reality, there were more than two mosaics. The case is discussed in detail in Part II, section "Mosaic Split and Mosaic Merge".

entific community at some historical period, then it would mean that there was no accepted mosaic at that period and therefore no scientific change. Consequently, there would be nothing at that time period for TSC to explain. Similarly, if HSC showed that at a certain time period there were two distinct coexisting communities each with its own distinct mosaic of theories and methods, then it would be a task of TSC to explain changes in both of those mosaics. In short, TSC should account for all those time periods when there existed a scientific mosaic.

The conclusion that can be drawn here is that, ideally, TSC should account for all scientific changes at all time periods. Of course, unearthing the mosaics of different communities at different time periods is an enormously challenging historical task. For many time periods and communities it may even prove insoluble. Yet the important point is that any scientific mosaic that HSC manages to reconstruct should be covered by TSC.

A related question arises regarding scientific fields (domains, disciplines): scientific change in what fields of science should TSC account for? Which scientific disciplines should be taken into consideration by TSC? Should it account for changes in natural science, or social science, or both? Should it explain transitions in formal sciences? Some of the existing theories were meant to account only for a limited subset of scientific fields. Kuhn's theory, for instance, was initially meant to apply only to the "mature" physical sciences (Although, on closer scrutiny it turns out that it is applicable to other fields as well). Other theories, such as that of Lakatos or both theories of Laudan, were intended as descriptions of scientific change in all domains.[157] In order to clarify the scope of TSC, we should find out whether it should be applicable to all or some fields.

The answer to this question is similar to that of the previous one. It follows directly from the definitions of scientific change and TSC that TSC should account for all changes in the scientific mosaic, regardless of which fields of inquiry they concern. Whether the scientific community replaces one physical theory with another, or substitutes one biological theory for another, or gives up one sociological theory in order to accept another, or replaces one accepted historical hypothesis with another – all these are instances of scientific change to be explained by TSC. A scientific change is to be explained even if it concerns a transition from one accepted theological or astrological theory to another at those time periods when these fields were considered part of the scientific mosaic. If a theory is considered accepted in the mosaic, both transitions to and from it should be accounted for by TSC.

Consider the place of theology in the Aristotelian-medieval mosaic. It was almost a commonplace for the modern historical accounts to present medieval theology as something completely opposed to genuine science. The picture created during the enlightenment and maintained until recently was that prior to the Scientific Revolution there was dogmatic and extremely unscientific theology on the one hand and a few "heroic" attempts of engaging in real science on the other; the dominance of theology was considered a serious obstacle to the progress of science. This modernist position is nowadays considered extremely anachronistic and rightly so. It

---

[157] See Laudan et al. (1986), pp. 159–160 and Nickles (1986), p. 254.

ignores the historical fact that, in the 1500s and 1600s theology was not something foreign to natural philosophy. Quite the contrary: the mosaic of the time included both theological and natural-philosophical propositions. Such theological propositions as "God is the prime mover" or "there is a strict distinction between the divine and the non-divine" were part of the mosaic of the time on a par with propositions of natural philosophy and cosmology such as "there is an absolute contrast between celestial and terrestrial regions" or "the motion of the celestial spheres is a cause of the variety of change in the sublunar (terrestrial) region". Moreover, not only were these propositions accepted together, but they were strongly interconnected. Take for instance, the proposition that the lowest regions of the universe are corrupted, alterable, and mortal, whereas the highest regions are pure, unalterable, and immortal. This cosmological proposition is immediately deducible from the conjunction of the two following theological propositions – "divine is immortal, unalterable, pure, and highest; non-divine is mortal, alterable, corrupted, and lowest" and "the structure of the universe reflects the divine/non-divine distinction". It is safe to say that, within the Aristotelian-medieval mosaic, theology and natural philosophy were intimately linked. Luckily, all this seems to be perfectly understood nowadays.[158]

Or take another example. The view that the world is created by God and that God occasionally intervenes in the world was part of the Aristotelian-medieval mosaic not only in the fifteenth century but even in the late seventeenth century. This is, essentially, the view that we nowadays call theism. However, by the early nineteenth century, at least in some communities, the accepted view becomes that of deism, which shares with theism the creation thesis, but denies God's intervention.[159] Finally, nowadays the scientific community seems to accept the agnostic view, according to which, it is no business of science to speculate upon such matters as the existence and nature of God. What we have here are two instances of scientific change – from theism to deism and then from deism to agnosticism. Although it is for specific historical research to establish when and under what circumstances these changes occurred in different mosaics, one thing is clear: there was a time when theological propositions used to be part of the mosaic.[160]

It must be emphasized that for TSC it makes no difference what field of inquiry a particular proposition pertains, as long as the proposition is part of the scientific mosaic, i.e. as long as it is accepted by the scientific community as the best available description of its object. After all, disciplinary boundaries are neither clear-cut nor fixed.[161]

---

[158] See Kuhn (1957), pp. 91–92. On the relationship between medieval theology and natural philosophy, see Lindberg (2008), pp. 228–253; Grant (2004), pp. 165–224.

[159] See, for instance, Olson (2004), pp. 124–131. Obviously, this historical hypothesis is in need of a detailed study.

[160] Brown has pointed this out in his (2001), pp. 107–108.

[161] Brown has mentioned this in (2001), p. 131. Apparently, this view goes back to Leibniz. See Yeo (2003), p. 243 and references therein.

First, disciplinary boundaries are not clear-cut, since at any given point of time there have been accepted propositions which pertained to more than one discipline. Take, for instance, our conviction that all material bodies are made of molecules. Does this proposition pertain to physics or chemistry? This vagueness of boundaries is not exclusively a modern phenomenon. As we have seen, the Aristotelian-medieval dichotomy between celestial and terrestrial was not only nature-philosophical but also theological, for it assumed that this dichotomy reflected the dichotomy between the divine and the non-divine. The fuzziness of disciplinary boundaries is a historical fact and it has a very simple explanation: the objects of different disciplines do not exist in complete isolation from one another.

In addition, disciplinary boundaries are evidently transient. Many propositions that we currently consider cosmological or physical would be taken as part of natural philosophy some 300 hundred years ago. Similarly, medieval optics was not merely a study of light but also included what we nowadays would call physiology of vision. Naturally, when the contemporary structure of disciplines and their sub-divisions is applied to mosaics of the past, it inevitably leads to anachronistic distortions. For instance, what would happen, if we were to apply our contemporary demarcation between science and theology to the seventeenth to eighteenth century mosaic? The consequences would be disastrous, for we would have to erroneously exclude theology from the scientific mosaic of the time, while in reality theology was part of the mosaic and not something foreign to it. We would thus anachronistically distort the real picture, by engaging in what is known as "tunnel history".[162] One way to avoid this is to accept that, for TSC, it makes no difference what particular discipline a proposition comes from as long as it is part of the scientific mosaic, i.e. as long as it is accepted by the scientific community of the time.

Although this answer seems trivial, there is more to the story. It is a historical fact that not all fields of science have managed to produce full-fledged theories accepted by the scientific community. There are fields of inquiry which are yet to produce theories as elaborate as those of, say, physics or chemistry. Take, for example, cultural studies or political science. As noted by many critics, most of what is sold under the label of "cultural studies" are works of literary criticism rather than full-fledged theories.[163] Similar accusations have been made against political science. It has been pointed out that "political science is still at a primitive stage, particularly on the theoretical side; so much so that much of it consists in commentaries on the classics and analogies are sometimes passed off as theories."[164] In short, it has been argued that in such fields as cultural studies or political science we can hardly speak of any currently accepted theories.

But the actual situation is more subtle. It is true that in such disciplines as cultural studies, political science, and even sociology there are virtually no accepted theories which could compare in their scope and refinement with theories of physics or biology. However, it is safe to say that there are many accepted general proposi-

---

[162] See Wilson and Ashplant (1988b), pp. 264–265.

[163] See Bunge (1998), p. 220.

[164] Bunge (1998), p. 157.

tions about the workings of society, culture, and politics. Consider such generalizations as "there are specific norms of behaviour in all societies", "all social systems deteriorate unless repaired", "rapid population growth causes overcultivation and deforestation", "war stimulates technical invention", "in stratified societies the dominant values are those of the ruling class(es)", "normally, critical thinking flourishes under political freedom and withers under tyranny", "during famine most people do not lie down, but scavenge for food, beg, steal, loot or move elsewhere".[165] That these propositions are accepted by the scientific community is apparent from the fact that they are often taken as truisms and only very seldom formulated explicitly. It is this feature of sociological generalizations that often prevents us from seeing that there is some content in them. But it is obvious that these generalizations are not self-evident (i.e. they do not hold in all possible worlds), for their opposites are not self-contradictory. And as long as we believe that they describe social processes correctly, they should be considered accepted by the community and, therefore, part of the scientific mosaic.

While the existence of accepted general propositions of the social sciences is beyond question, there is currently no agreement whether there are or can be any genuine social laws. To be sure, many of the accepted general propositions of social science have been labelled "laws". Consider, for instance, the so-called Duverger's law of political science, which states that political systems that feature simple-majority single-ballot electoral systems tend to evolve into two-party systems. According to Duverger, in single-member simple plurality districts, voters generally realize that if they vote for a third-party candidate, they will likely be throwing their vote away, for the third-party candidate is not a serious contender. Thus, they often vote for one of the two leading-party candidates. As a consequence, gradually a two-party system emerges.[166] Now, it has been argued that Duverger's law and other social laws are mere empirical/statistical generalizations which do not qualify as genuine laws, since they allow for exceptions. Duverger's law, for instance, faces many exceptions, for third parties do sometimes manage to succeed even in simple-majority single-ballot systems. The same holds for other so-called laws of the social sciences. It is safe to say that so far virtually all accepted social laws are riddled with exceptions.

Yet, the fact that none of the extant social laws is exception-less does not necessarily imply that there can never exist any strict exception-less social laws (akin to those of physics), i.e. it does not necessarily mean that all general propositions in social science are mere exception-ridden statistical/phenomenological generalizations. This is still an open question. While some authors claim that general propositions of social sciences are inevitably much weaker and less reliable than those of physics or biology, others argue that, in principle, there can exist strict social laws similar to those of natural science.[167] There is also the third (often ignored) possibil-

---

[165] See Bunge (1998), pp. 26, 28, 221, 280.

[166] See Duverger (1954), p. 217. For discussion, see Amorim Neto and Cox (1997).

[167] See Hempel (1942); Huntington (1968); Salmon (1989); Kincaid (1990, 2004); Little (1993); King et al. (1994); McIntyre (1993, 1998). For discussion, see Gorton (2014) and references therein.

ity that there can be exception-less social laws which are nevertheless probabilistic in nature, akin to the law of radioactive decay or the laws of Asimov's psychohistory. At the moment, we can't really predict whether social science will eventually succeed in constructing and accepting "genuine social laws" or whether it is destined to always deal with "mere phenomenological generalizations". In any event, the outcome of this debate is not crucial from the perspective of TSC. Regardless of the outcome, the task of TSC will remain intact, for it is clear that there are accepted general propositions in social science and acceptance/rejection of those propositions is clearly within the scope of TSC.[168]

Thus, it would be incorrect to say that such fields as sociology or cultural anthropology are not represented in the mosaic at all. In reality, while some fields are represented in the scientific mosaic by highly elaborate comprehensive theories, others have so far provided only separate loosely connected generalizations.

This also applies to philosophical conceptions. It is common wisdom that in philosophy there are no accepted theories whatsoever. This wisdom, however, is at best imprecise. Despite all the controversies, there have been and there still are many accepted philosophical propositions. Take, for instance, the conception of infallibilism, according to which, infallible, demonstratively true knowledge about the world is attainable. It is safe to say that this view was part of the accepted mosaic for the most part of the history of knowledge. It is equally safe to say that it is no longer in the mosaic, for despite all our disagreements we nowadays accept the opposite position – that of fallibilism – which roughly holds that no contingent proposition can be demonstratively true and, thus, no theory in empirical science can be absolutely true.[169]

In addition, some disciplines are represented in the mosaic exclusively by singular propositions. This is true for all historical disciplines such as military history, social history, or HSC. Naturally, we do not expect our historical disciplines to make generalizations about the past. What we do expect is correct descriptions of individual events. Clearly, such propositions as "Augustus was the first emperor of the Roman Empire", "the Battle of Kursk took place in the vicinity of the city of Kursk in July and August 1943", "Isaac Newton was born in 1642", or "Albert Einstein received the 1921 Nobel Prize in Physics" are part of the contemporary mosaic, for we accept them as correct descriptions of the events of the past. Thus, changes in our historical knowledge are as much of interest as transitions from one accepted physical theory to the next.

Finally, it may even turn out that the mosaic contains ethical beliefs, such as the contemporary belief in equality of rights or the principle of tolerance. This is however a questionable point, for at this stage we cannot say whether the mosaic does in fact contain such normative propositions, as "racial discrimination is immoral" or

---

[168] Since the debate is still open, we can use the term "law" as it has been traditionally used in social science, i.e. without committing to any of the opposing views on the nature of the so-called social laws.

[169] For details, see Lakatos (1970) and Laudan (1980).

"one ought to help those in need"; this is still very much an open question that TSC and HSC must tackle together. What seems clear though is that ethical beliefs do sometimes find their way into the mosaic when they are construed as descriptive propositions. Take, for instance, the medieval ideal of a warrior king, according to which "the man waging for the common good of the realm, was also the one most worthy of the crown of the kingdom".[170] While nowadays we would declare this proposition normative ("a proper king ought to fight for the good of the realm"), back in the Middle Ages it was taken as a true description of the very nature of kingship. The idea of fighting for the good of the realm was implicit in the accepted concept of kingship: for the community of the time, to be a king meant, among other things, to be ready to fight for the good of the realm. The medieval principle of hierarchy is another example. The principle states: "all forms of things are ranked in a natural hierarchy according to their degree of perfection". It was believed that all things exist within a universal hierarchy that stretches from God at its highest point to inanimate matter at its lowest. This principle was part of the conception of the Great Chain of Being accepted in the Middle Ages.[171] For the medieval community this wasn't merely a normative proposition prescribing how things ought to be; first and foremost this principle was understood as a description of the actual state of affairs, where kings are naturally more perfect than bishops, bishops are more perfect than knights etc. Thus, when it comes to descriptively formulated views on ethical and social issues, they can be part of the mosaic. Once we appreciate this, it becomes clear that at least part of what the sociologist of science would include in the illusive "sociocultural context" is just a set of accepted descriptive propositions on ethical, social, and cultural issues. Such propositions are part of the scientific mosaic and, thus, are within the scope of TSC. What remains to be settled is whether normative propositions such as those of ethics or methodology can also be part of the mosaic. This calls for a careful investigation. If it turns out that normative propositions can also be part of the mosaic, then their acceptance and rejection will be within the scope of TSC.

In short, we should not expect every field of inquiry to be represented in the mosaic by complex theories with hundreds or thousands of propositions. Some fields may partake in the mosaic by only a few accepted propositions. These propositions may or may not be systematically linked; it is possible that they are interconnected only loosely, or even not at all. From the metatheoretical perspective, it makes no difference whether we deal with an elaborate system of propositions or an individual proposition. The task of TSC remains intact: it should explain scientific change, be it a transition from one complex theory to another or from one individual proposition to another. TSC should explain all transitions in the mosaic, regardless of which fields of inquiry the transition concerns.

This brings us to the final question of this section which concerns the scale of scientific changes that TSC is to explain. Should TSC deal with only grand changes such as the transition from the Aristotelian-medieval natural philosophy to that of

---

[170] Kantorowicz (1957), pp. 259–260.

[171] The idea of the Great Chain of Being is scrutinized in Lovejoy (1936).

Descartes and his followers, or should it also be concerned with minor changes such as the transition from Newton's original theory of Earth's shape to that of Alexis Clairaut (circa the late 1740s), or from Newton's lunar theory to that of Tobias Mayer (circa the mid-1760s)?

Some authors have insisted on the necessity of differentiating between two types of change. In Lakatos's conception, for instance, there is a clear-cut distinction between transitions within a specific series of theories (e.g. from one version of Newtonian physics to the next) and shifts from one research programme to another (e.g. from the Cartesian natural philosophy to that of Newton). While the transitions of the former kind have to do with the so-called protective belt of auxiliary hypotheses, the transitions of the latter kind concern the very hard core of a research programme. An analogous dichotomy is present in Laudan's early theory, where he distinguishes between theories and research traditions. Similarly, in his Treatise, Mario Bunge discriminates between specific theory (theoretical model) and generic theory (framework). In Kuhn's model, a distinction is drawn between transitions during the period of normal science and revolutionary transitions which concern whole paradigms. Finally, a two-process model of scientific change is also assumed in the VPI project. The members of the project distinguish between what can be called grand changes (those of guiding assumptions) and minor changes (those of actual theories within the framework of guiding assumptions).[172] As different as they are, all these distinctions have a common trait – they all assume that there are two different types of scientific change, each with its own peculiarities.

The alternative position is that scientific change is a unitary process. Popper's conception is a vivid example of this. According to Popper, the mechanism of scientific change is essentially the same regardless of the calibre of change. In essence it is a series of conjectures and refutations (or attempts to refute proposed hypotheses to be more precise). Whether we deal with the transition from the Newtonian to the Einsteinian theory of gravity, or from one version of wave optics to another, or from one lunar theory to another, scientific change, according to Popper, has the same logic. Quine agrees with Popper on this point. In Quine's conception, any scientific change, be it great or small, is always a change in our overall web of belief. It is essentially an attempt to incorporate some new evidence into the web of belief.[173]

Whether scientific change involves two distinct processes or whether it is a unitary process is a question that refers to the object-level. Therefore, it should be left to an actual TSC to settle. The metatheoretical question that concerns us here is whether all kinds of scientific change should be explained by TSC, or whether there are spe-

---

[172] See Kuhn (1962/70); Lakatos (1970), pp. 33, 41, 48; Laudan (1977), pp. 70–120; Bunge (1974), pp. 99–101; Donovan et al. (eds.) (1992). For discussion, see Godfrey-Smith (2003), pp. 117–121.

[173] See Quine and Ullian (1978).

cific kinds of change which lie beyond the scope of TSC. As in the case of the two previous questions addressed in this section, the answer to this question too immediately follows from the definitions of scientific change and TSC: all changes should be accounted for by TSC.[174] In fact, both those who agree that scientific change is a unified process and those who deny it normally hold that scientific change should be explained irrespective of what particular category of change it falls into.

What makes this question interesting from the metatheoretical perspective, however, is that sometimes in actual practice we seem to forget that all changes are subject to explanation. Take, for example, the popular portrayals of the famous Duhem-Quine thesis. It is often presented as stating that one can save a theory from empirical refutation by making necessary adjustments and modifications in some of the auxiliary hypotheses. This may create a false impression as if choice of auxiliaries were something completely arbitrary, as though it were quite legitimate to say "What, has the prediction of a theory failed? Oh, no worries, I'll change the auxiliaries here and there and it'll be just fine!".[175] But it is obvious that the scientific mosaic is not tampered with in that way. Auxiliary or not, a proposition that is accepted by the community constitutes part of the mosaic and, thus, is treated accordingly. A suggested replacement of an accepted auxiliary is an attempt to modify the mosaic and, therefore, is to be accounted for by TSC.

Consider another example. It is well known that one of the novel predictions of general relativity was the phenomenon of light-bending. Moreover, as the historical record shows, it is only after this prediction was confirmed by Eddington that general relativity became accepted. However, it has been argued that the Newtonian theory could also predict the gravitational deflection of light provided that the auxiliary assumptions were picked so as to produce the desired results. Thus, in order to predict the phenomena of gravitational deflection of light by means of the Newtonian theory one must "only" dismiss the theory of electromagnetism and replace it by a version of the corpuscular theory of light.[176] But such a portrayal creates a false impression as if the choice of auxiliaries were completely arbitrary. It takes theories out of their historical context by disregarding the state of the scientific

---

[174] Barberousse has presented the following dilemma: either (1) we delve into the details of each minute modification and risk making scientific change intractably small or (2) we focus only on general formulations of theories and grand transitions in order to make the process intelligible. See Barberousse (2008), p. 88. This is obviously a false dilemma, since we have to trace changes in the mosaic regardless of their scale. The fact that this might be quite a challenging task is not an excuse for a retreat.

[175] It should be noted that the proponents of two-process models of theory change can by no means be held responsible for this misunderstanding, for they make it perfectly clear that introduction of auxiliaries should also be appraised. Moreover, Lakatos's whole methodology (his (1970)) can be taken as an attempt to differentiate between progressive and regressive modifications of auxiliaries.

[176] See, for instance, Jaki (1978) and Will (1988).

mosaic of the time. In reality, there is always the scientific mosaic in a particular state and respective attempts to modify it. Indeed, when in 1801 Johann Georg von Soldner predicted the gravitational deflection of light, he based his calculations on the then-accepted corpuscular theory of light. This prediction followed from the mosaic of the then-accepted theories and continued to follow from the mosaic up until the acceptance of the wave theory of light in the 1820s. But to claim that the same prediction followed from the Newtonian theory of gravity at the time when general relativity was proposed is a clear-cut instance of anachronism, since in the 1910s the scientific mosaic included (among an array of many other theories) a version of the Newtonian theory of gravity and the Maxwellian electrodynamics together with the wave theory of light. The corpuscular theory of light was not in the mosaic of the time. Therefore, the phenomenon of gravitational deflection could not be predicted from the propositions of the scientific mosaic of the mid-1910s. It was not until after general relativity was accepted that the prediction of gravitational light deflection could be obtained from the scientific mosaic.

Such anachronisms can be avoided if we keep in mind the state of the mosaic of the time period under scrutiny and appreciate that the so-called auxiliaries belong to the mosaic as much as anything else and that they are not chosen in an arbitrary fashion. In short, from the metatheoretical standpoint, it is obvious that TSC should explain all modifications of the scientific mosaic: whether grand or minor, any replacement of any element of the mosaic is an instance of scientific change and is to be accounted for by TSC.[177]

To be sure, an actual TSC may distinguish between grand and minor changes, or between scientific revolutions and normal-science changes, or between hard core and auxiliary changes, or between some other genera of change – there is nothing wrong with that. But it should necessarily provide explanations for all kinds of change that the mosaic undergoes.

To sum up this section, any scientific change is to be accounted for by TSC, irrespective of how grand or how minor it is, regardless of what time period it belongs to, or what scientific field it concerns. Ideally, TSC should account for all transitions from one theory to another or from one method to another.

Epistemology, History and Theory of Scientific Change

So far, my task has been to clarify what TSC is and, importantly, what it is not. First, as I have attempted to show, TSC is not to prescribe any methods of theory appraisal; it shouldn't be confused with normative MTD. Also, TSC doesn't necessarily have to explain how theories are constructed (created, developed), how they become

---

[177] For discussion of the case, see Leplin (1997), pp. 78–79. While Leplin clearly realizes that the choice of auxiliaries is not arbitrary, other popular accounts of this historical episode commit a fatal error. Bertrand Russell, for instance, presents the case as though it were completely admissible to switch from the wave theory of light to the corpuscular theory. See Russell (1959), p. 130.

useful or what makes them pursuit-worthy. Finally, TSC is not a study of changes in individual belief systems; while an actual TSC may turn out to be applicable to the individual level, this is not the purpose of TSC. What TSC should do is account for each and every change in the mosaic of accepted theories and employed methods, regardless of its time period, scientific field, or scale. Now that we know the scope of TSC, we can position TSC among its siblings – history of scientific change (HSC) and epistemology (EPI).

Before considering the relation between TSC and HSC, a short historical note is in order. When the field of the history and philosophy of science (HPS) was born in the 1960s, its main rationale was that we could study the actual workings of science in order to use that knowledge to answer some key questions of the epistemology and methodology of science, such as "what is the demarcation between science and non-science?", "what makes one theory better than another?", "do scientific theories provide true/truthlike descriptions of the mind independent world?", and "is there scientific progress?". Kuhn and other founding fathers of HPS suggested that in order to solve the philosophical problems of scientific rationality, progress, realism, and demarcation we had to study the actual historical episodes. We would then rely on the findings of philosophy of science to make sense of individual historical episodes. This idea was nicely summed up in Lakatos's famous dictum that "philosophy of science without history of science is empty; history of science without philosophy of science is blind."[178] This wasn't merely Lakatos's own ideal; it was shared by many of his contemporaries. It is safe to say that this classic HPS is now well and truly dead.

As we have seen, the theories of the classics had many fatal flaws. One of the flaws implicit in many of the theories produced in the 1960s and 1970s was the assumption that the core method of science is universal and unchangeable. Popper, Lakatos, the early Laudan and many of their colleagues tried to explicate this universal and unchangeable method.[179] In order to show that their own explications were correct, they would apply their methodological conceptions to different historical episodes and, as a result, would often end up shoehorning these episodes into their methodological schemes.[180] It is now obvious that the whole enterprise was doomed since, as we know, methods of science are changeable. They were looking for a black cat in a dark room, while there was no cat. Consequently, the whole practice gradually withered and nowadays HPS is no more than an umbrella term; historians and philosophers pursue their separate projects with essentially very little overlap.[181]

---

[178] See Lakatos (1971), p. 102.

[179] See section "Descriptive and Normative" above for discussion.

[180] Some of the most famous examples are Agassi (1972), Lakatos and Zahar (1976), Zahar (1973). Unsurprisingly, this malicious practice was harshly criticized. See Williams (1975). In Kuhn's own words, "what Lakatos conceives as history is not history at all but philosophy fabricating examples" (Kuhn 1970b), p. 143.

[181] There are several historical accounts of the fate of HPS. See for instance Nickles (1995), Laudan (1990).

Meanwhile, during the last 30 years, there has been a fascinating growth in the field of the history of science. Historians have produced a variety of different approaches and greatly enhanced our knowledge of historical episodes. Based on the object of study, the historian can focus (1) on individual scientists, (2) on communities of scientists and (3) on the product of science – scientific theories and methods. It is safe to say that most of the history of science nowadays focuses on individual actors and their sociocultural contexts, while the study of the scientific mosaic seems to be of secondary interest, but it is the latter we are mostly interested in here, since the history of scientific change (HSC) is precisely that branch of the history of science that concerns changes in the scientific mosaic.

Nowadays, histories of theories and methods are usually written without any openly accepted guiding theory,[182] as if theory-ladenness never existed. However, while the openly prescribed methodology of HSC tells us to stay away from general propositions, the actual method employed in HSC certainly allows for generalizations (albeit only tacit). It is of course obvious from the phenomenon of theory-ladenness that a purely non-theoretical science (i.e. a science without any general propositions) is an oxymoron. As Mario Bunge observed almost half a century ago in another context, those who opt to do without a systematic theory, usually end up implicitly accepting some ad hoc "home-spun" assumptions.[183] In short, no historical account can proceed without general propositions – explicit or tacit – and HSC is no exception.

Consequently, we are facing a dilemma: either we start formulating our general historical assumptions openly, or we continue keeping them tacit. The drawbacks of relying on tacit assumptions are obvious. If not explicitly stated, an assumption remains accepted uncritically and cannot be properly scrutinized and changed. In addition, when general assumptions are left tacit, different narratives can rely on mutually incompatible assumptions; this can easily result in a fragmentation of the field. Finally, if an argument is based on tacit assumptions, evaluating its validity and soundness becomes problematic. Thus, the preferable strategy is to formulate our general propositions openly, construct a systematic general TSC, and cease the practice of relying on tacit assumptions. This will lead to a mutually beneficial cooperation of HSC and TSC.

In the model I am proposing here, HSC should test general hypotheses about the process of scientific change and provide TSC with necessary historical data, while TSC should provide HSC with questions that need to be answered, a systematic vocabulary, explanatory tools, and suggest which features are relevant to the process of scientific change:

---

[182] There are, of course, notable exceptions such as Chang (2004).

[183] See Bunge (1973), p. 1. This is a well-known phenomenon. See also Donovan et al. (eds.) (1992), p. xviii.

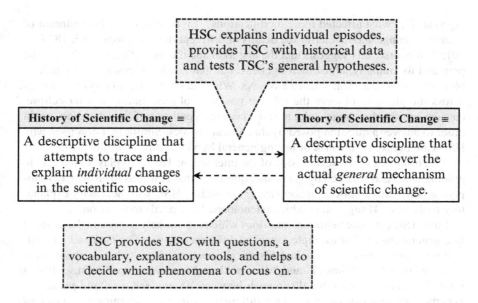

HSC explains individual episodes, provides TSC with historical data and tests TSC's general hypotheses.

History of Scientific Change ≡

A descriptive discipline that attempts to trace and explain *individual* changes in the scientific mosaic.

Theory of Scientific Change ≡

A descriptive discipline that attempts to uncover the actual *general* mechanism of scientific change.

TSC provides HSC with questions, a vocabulary, explanatory tools, and helps to decide which phenomena to focus on.

The first side of the above interaction is relatively straightforward: it is obvious that only HSC can apply the general propositions of TSC to actual historical episodes, only HSC can test general hypotheses concerning the mechanism of scientific change proposed by TSC and only HSC can provide TSC with relevant historical data.

There has been a question whether the historical narratives, as produced by contemporary HSC, are of any use for TSC. Pinnick and Gale, for instance, have argued that most of the contemporary historical narratives are of little use as far as TSC is concerned.[184] They are pessimistic about the prospects of using the fruits of contemporary HSC for the purposes of TSC. Moreover, they think that HSC is, in principle, incapable of providing the data sought by TSC, for HSC is inherently narrativist and, consequently, inevitably *particularist*.[185] Their proposal is to devise an alternative "philosophical" approach to the history of science which will consider historical cases from the philosopher's angle, paying attention to the general features of scientific change.

Of course, it is true that our contemporary narratives do not always appear readily useful for TSC, since they often have a focus quite different from that of TSC.[186] However, I am optimistic that HSC can do the job itself; no alternative "philosophical historiography" is necessary. There is nothing preventing HSC from asking questions about the scientific mosaic, its successive states, and corresponding changes. It is the task of HSC to answer questions devised by TSC such as "what was the state of the mosaic at time *t*?", "what modifications took place in the follow-

---

[184] Pinnick and Gale (2000), p. 115. Naturally, they follow the tradition and say "philosophy", not "TSC".

[185] Pinnick and Gale (2000), p. 110.

[186] See section "Relevant Facts" below for discussion.

ing years?", "what affected these modifications?". Who else if not the historian of scientific change should clarify the details of different mosaics circa 1615, 1815, or 2015? Of course, this suggests that the historian and the theoretician will have to be prepared to mutually adjust their agendas, but that is what happens in any pair of observational-theoretical science anyways. When an observational physicist travels across the globe to observe the relative positions of stars during a solar eclipse, chances are she does so in order to test a certain hypothesis of light-bending, i.e. in order to answer a question posed by theoretical physics. Similarly, I don't see why HSC cannot take on the task of testing general hypotheses proposed by TSC.

This brings us to the second side of the interaction: TSC is to play a vital role in guiding historical research by (1) posing questions that HSC has to answer, (2) supplying HSC with a uniform taxonomy, (3) providing HSC with necessary explanatory tools, and (4) suggesting which phenomena HSC needs to focus on.

First, TSC can pose historical questions which are unlikely to arise in the absence of a general theory. For example, once the difference between *method* and *methodology* is appreciated, two distinct historical question arise: "what methods were actually employed at time *t*?" and "what methodologies were openly prescribed at time *t*?". Similarly, once we distinguish between *acceptance*, *use*, and *pursuit*, it becomes obvious that there are three different questions concerning the stance of the community: "was theory *x* accepted at time *t*?", "was theory *x* used at time *t*?", "was theory *x* pursued at time *t*?". TSC can reveal that there is a vast and barely touched layer of history – that of the transitions from one state of the scientific mosaic to the next. Indeed, so many interesting historical questions remain not only unanswered but even unasked in the absence of an appropriate guiding theory.[187] As we shall see in *Part II*, an actual TSC poses many more historical questions, which would simply be unthinkable in the absence of a guiding theory.

Secondly, TSC may provide HSC with a uniform systematic *taxonomy*. This will help to prevent a fragmentation of historical narratives and will potentially open the doors for creating a unified historical database. Once the key theoretical concepts are in place, practicing historians will be able to present their historical explanations by using the same vocabulary and, if necessary, add records in respective tables of a unified historical database. There is a challenging analytic task of designing such a database. Clearly, no such database is possible in the absence of a universal taxonomy, which is itself impossible in the absence of an accepted general theory.

Thirdly, once constructed and accepted, general propositions of TSC will play an important role in *explaining* individual historical episodes. Instead of relying on tacit "common sense" generalizations, HSC will be in a position to *explicitly* refer to this or that general proposition concerning the mechanism of scientific change. Suppose, the historian tries to find out why the discovery of the famous anomaly of Mercury's perihelion in 1859 did not result in the rejection of the then-accepted Newtonian theory of gravity. When studying the episode, she may tacitly assume that, in general, empirical anomalies are not fatal to accepted theories. A number of questions arise concerning this tacit assumption. Is it indeed

---

[187] See Pinnick and Gale (2000), p. 118 for a similar view.

true that empirical anomalies are not fatal to accepted theories in all fields of inquiry and at all times? Or is it possible that our attitude towards anomalies is changeable? If so, then how and why does our attitude towards anomalies change? Does it change randomly, or does it change in a rational, law-governed fashion? In order to answer these questions, we will have to formulate our general propositions explicitly and construct a systematic TSC. If an actual TSC manages to unearth the mechanism of scientific change and becomes accepted by the community, it will act as an explanatory framework for HSC; from then on, the historian will no longer have to rely on tacit "common sense" assumptions about the process of scientific change, but will explicitly refer to the accepted propositions of TSC as premises of her historical explanations.[188]

Finally, TSC can help to differentiate between relevant and irrelevant facts of the history. Historians agree that "one of the most challenging problems in writing history is the selection of which details to include (or exclude)."[189] It is obvious that the historian cannot possibly focus on all the features of a period under scrutiny; she must choose which features to discuss and which to ignore. In the absence of a guiding theory, the historian has no other choice but to rely on some tacit "common sense" criteria of relevance, which is obviously far from ideal. A guiding TSC can certainly help to solve this problem, by suggesting which features of the process must be recorded by the historian, just as theoretical physics suggests which observable features of a physical process to pay attention to. For example, if the accepted theory suggests that the motion of planets is governed by the Newtonian laws, the observer knows which types of parameters to focus on (e.g. mass, velocity, distance, acceleration) and which to disregard (e.g. size, internal chemical structure, atmosphere composition). Similarly, if an actual TSC suggests that the employed criteria of theory assessment change as a result of changes in underlying assumptions about the world, then unearthing both the criteria and the assumptions on which they are based becomes an important task for HSC.[190]

---

[188] There is an ongoing debate on the nature of explanation in historical sciences. See footnote 164 above for references.

Regardless of which of the extant views on explanation in the social sciences turns out to be correct, one thing is clear: any proper explanation needs to refer to some general propositions, be these "proper laws" or "mere generalizations". Even those who deny the possibility of genuine laws in social science, accept that some general propositions (albeit not law-like) are inevitable. See Elster (2007), pp. 36–37; Glennan (2010).

[189] Maienschein and Smith (2008), p. 320.

[190] It is readily seen that this list of the functions of TSC is not exactly what Lakatos had in mind. I agree with Lakatos that any result of observation or experiment is essentially theory-laden and historical "facts" are no exception. There can be no "pure" (non-theory-laden) facts of HSC, just as there can be no "pure" observational propositions in physics. This much is clear. Yet, the key difference between my position and that of Lakatos is that, in my view, HSC is laden with a *descriptive* TSC, whereas Lakatos claims that "history of science is a history of events which are selected and interpreted in a *normative* way" (Lakatos (1971), p. 121, my emphasis). As I have explained in section "Descriptive and Normative", this difference is crucial for understanding the scope of *descriptive* TSC.

In brief, it is obvious that both TSC and HSC would mutually benefit from this collaboration. While HSC would test general hypotheses proposed by TSC and would provide TSC with required historical data, TSC would provide HSC with a uniform taxonomy, pose important questions, offer necessary explanatory tools, and suggest which phenomena to focus on. Such an arrangement would help us avoid two vices. On the one hand, we would avoid fabricating whiggish histories, as we wouldn't be trying to shoehorn historical episodes into the schemes of this or that normative MTD. On the other hand, we would also avoid the other extreme of uncritically accepting tacit general assumptions. We would instead engage in a fruitful collaboration where (1) new TSC's will be constantly built and modified based on the findings of HSC, (2) these new theories will be constantly scrutinized by HSC and (3) the accepted TSC will be used by HSC to explain historical episodes.

Let us now turn to the relation between TSC and HSC on the one hand and EPI on the other. There is an unfortunate tendency nowadays to forget that the initial motivation for studying the actual process of scientific change was the idea that several important epistemological issues cannot be settled unless we refer to the history of science. The whole field of HPS originated from the conviction that once we understood the actual workings of science we could then use that knowledge as a means for answering important questions of epistemology such as "is there growth of knowledge?", "is there scientific progress?", or "can our descriptions of the world be true, or approximately true, or probable?". This idea runs through the works of Kuhn, Lakatos, Laudan and many others. However, as I have explained in section "Descriptive and Normative", the founding fathers of HPS committed a serious crime by mixing up the problems of TSC, EPI, and MTD in a juicy but indigestible cocktail.

Consider a scenario that is quite common these days. The epistemologist wants to solve a particular epistemological issue. What does she do? Quite often the epistemologist understands that in order to back up her position she needs to refer to the actual workings of science. However, having no accepted theory of scientific change, she decides to fabricate her own theory of scientific change. Consequently, she is forced to show that her tailor-made theory is plausible. In order to do that, she starts searching for relevant historical data, but since the existing accounts of historical episodes do not necessarily support the newly tailored theory, she ends up engaging in original historical research. The result is a mishmash of EPI, TSC, and HSC.

Illustrations of this phenomenon are numerous. Take, for instance, Otávio Bueno's recent paper where, in order to argue for his *epistemological* conception (a version of van Fraassen's constructive empiricism, to be precise), he sets off to "sketch an alternative account of scientific change".[191] This is as though a medical

---

[191] Bueno (2008), p. 213.

practitioner wanted to find out what drug to prescribe, but, because of the absence of any tested medication and even any knowledge of human anatomy, ended up constructing brand new anatomical and medical theories instead. Or take Ronald Giere who finds himself discussing the *epistemological* issue of relativism in the same paper in which he outlines his *theory of scientific change*.[192] This is as though an engineer wanted to build a bridge but, for lack of technology or any underling theory, ended up tailoring his own material science and proposing new physical theories.

The only excuse for this futile practice is the fact that nowadays there is no accepted TSC. As a result, contemporary epistemologists of science are basically forced to tailor their own sketchy accounts of scientific change along the way. We desperately need but still lack an accepted TSC; the vast historical scholarship notwithstanding, our knowledge of the actual mechanisms of scientific change is extremely fragmentary and insufficient for the purpose. This often forces contemporary epistemologists to fill the gap by producing quasi-theories of scientific change, tailored to fit their particular epistemological preferences. "Theories of scientific change" devised by instrumentalists often ignore *acceptance* and focus exclusively on *use*,[193] while those produced by pragmatists usually cover only *pursuit*.[194] It is as though the cosmologist decided to devise her own quantum theory in order to make her favourite cosmological hypothesis plausible. Albeit daring and even heroic, this longing to solve all problems at once, I believe, is extremely ineffective. It is because of this hasty and unsystematic approach that presently there is neither an accepted TSC, nor any serious advancement in settling the respective epistemological issues.

What we need, therefore, is a *piecemeal* approach. We should admit that it is one thing to construct a TSC and assess it by HSC and it is another thing to discuss epistemological issues of growth of knowledge, progress, demarcation, realism etc. The outcome of the TSC-HSC partnership should be taken as a starting point for an EPI investigation; our epistemological preferences (e.g. realism/instrumentalism) should not be implanted into our general TSC from the outset. Of course, when a TSC becomes accepted, it will be quite useful in settling the problems of EPI, but this should not be taken as an opportunity for blending the two together.

To sum up, both EPI and HSC need an accepted TSC, while we are yet to construct one. Only when such a theory is constructed and accepted by the community, can we legitimately base our epistemological and historical discussions on our knowledge about the mechanism of scientific change. That is why the construction of a TSC is a task of utmost importance.

---

[192] See Giere (1984), p. 14.

[193] van Fraassen (1980) and Cartwright (1999) are good examples.

[194] What Feyerabend (1975) says about proliferation of theories can, at best, apply only to *pursuit*.

Chapter 2
Possibility

The fact that TSC is desirable doesn't automatically make it possible. After all, there are many highly desirable things which, in light of our current knowledge, are simply unattainable. Similarly, it may yet turn out that no TSC is possible in light of the current state of the scientific mosaic. In that hypothetical case, we would be forced to abandon this project altogether, just as many contemporary authors would readily suggest. Very few authors, to put it mildly, are nowadays sympathetic to the idea of constructing a general theory that would provide an explanation of scientific change applicable to different fields and different time periods. This is the position that I call *particularism*. The opposite view may be called *generalism*. Therefore, the key question is:

? Can there be a general theory of scientific change?	
Yes	**No**
Generalism: A general TSC is possible.	**Particularism:** There can be no general TSC.

The particularist position is nowadays supported by a set of often repeated arguments. Some of the arguments are specific to the generalism/particularism controversy on the possibility of TSC, while others are more universal in nature and go back to the famous *nomothetic/idiographic* distinction and the debate on the possibility of general laws in the social sciences in general. At this stage, it is not clear whether there are genuine laws that govern social processes. Obviously, this is not an opportune place for rehashing the well-known general arguments for and against the nomothetic/idiographic distinction.[1] Nor is this an opportune place to discuss the

---

[1] For discussion of general arguments for and against the traditional nomothetic/idiographic distinction, see Bunge (1998, pp. 21–33, 241, 257–269). For the specific debate on the possibility of laws in social science, see McIntyre (1993, 1998), Little (1993), and Kincaid (2004).

© Springer International Publishing Switzerland 2015
H. Barseghyan, *The Laws of Scientific Change*, DOI 10.1007/978-3-319-17596-6_2

possibility of general laws in the social sciences. My task in this section is to discuss the main arguments which are specific to the debate on the possibility of TSC.[2]

The Argument from Changeability of Scientific Method

One of the most common arguments against the possibility of TSC is *from change-ability of scientific method*. It runs along these lines:

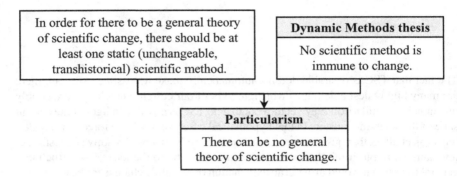

On the one hand, it is argued, if there is ever to be a general theory of scientific change there should exist the universal method of science, the set of unchangeable, fixed rules employed in theory assessment in all disciplines at all times. But, on the other hand, it is a historical fact that there is no such universal and fixed method of science, for we know that methods of science change through time. They change so drastically that there is nothing that could be said to be held in common between Aristotle and Einstein. Besides, even when we focus on a limited period of history, we are still unable to find one method common to all fields of science of that period. Thus, it is so concluded, the whole project of TSC is completely pointless. Both premises of this argument seem to be taken for granted nowadays.[3]

Although it is possible to doubt both of these premises, the question of whether there is a universal fixed method of science should be put aside for now, since it is not the task of the metatheory to touch upon this factual issue. Here we can only discuss the first premise, according to which, uncovering the general principles of scientific change amounts to explicating the method of science. It should be obvious

---

[2] Here I shall not repeat the arguments discussed in the context of the VPI project. For criticism, see Nickles (1986) and (1989), Preston (1994). For the reply of the members of the VPI project, see Donovan et al. (eds.) (1992, pp. xvi–xx).

[3] See Shapin (1996, pp. 3–4). The argument was also one of the central theses of Feyerabend. See, for instance, his (1975, p. xii). It was then taken up by Barnes and Bloor. See, for example, their (1982, p. 27). This same argument has been also levelled against the VPI project. See Nickles (1986) and Preston (1994, p. 1065).

from the preceding discussion that the premise stems from serious confusion. As I have shown in section "Descriptive and Normative", uncovering the laws that guide scientific change is not the same as explicating the method employed in theory appraisal. For it may well turn out that all methods are changeable and yet there are some laws that govern their change. And it is these laws that TSC should uncover. In other words, the task of TSC and HSC is to explain scientific change – if it turns out that methods also change, then TSC and HSC would have to explain changes in theories and methods alike. The task of TSC is to not to locate the alleged universal and unchangeable method of science, but understand the common mechanism that underlies transitions from one method to another. We have a very illustrative example of such a TSC – the later Laudan's *reticulated model* presented in his *Science and Values*. This reticulated model doesn't presuppose any unchangeable method of science, but attempts to explain transitions in methods and theories alike. The fact that Laudan's model doesn't succeed in explaining the process is irrelevant here; the important point is that no universal method needs to be presupposed in order for TSC to be possible. This voids the arguments *from changeability of methods*.

The Argument from Nothing Permanent

The argument from changeability has nowadays a more robust analogue, which we may call the argument *from nothing permanent*. It can be outlined thus:

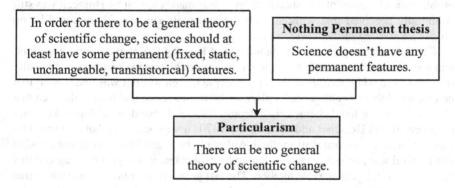

It has become fashionable lately to insist that no feature of science is transhistorical. Science, it is often argued, has no fixed (static, unchangeable) features whatsoever. But in order for there to be a general TSC, so it is purported, science should have at least some objectively transhistorical features. Therefore, the conclusion goes, no general TSC is possible. Note that this argument *from nothing permanent* avoids the confusion of TSC with MTD (the confusion that was implicit in the argument from *changeability of scientific method*).

I have no quarrel with the first premise of the argument. It is clear that no TSC is possible unless *some* objective features of science persist through all transformations. Even if no theory or method persists through the process of scientific change, at least the *laws* that govern the process must be permanent in order for TSC to be possible. Similarly, there could have been no physics had the world contained no regularities whatsoever, or had the regularities changed in random fashion. Generally speaking, a theory of something is possible only if that something has some objectively stable features. This much is clear.

My problem is with the second premise of the argument – *the nothing permanent thesis* – which suggests that not only methods of science but *each and every feature* of science is transient. Nowadays, many authors take *the nothing permanent thesis* as something established beyond any reasonable doubt. Take Galison for instance, who says that "the competitive campaign fought in the 1970s among philosophers to find *the* theory of scientific change no longer grabs philosophers of science as a plausible enterprise. Science seems far too heterogeneous for that: too diverse at a given time (especially now); even within the same subdiscipline too much has changed"[4]. A short historical note is in order here.

Long ago, in the "good old days" of justificationism, it was accepted that the scientific mosaic included at least some permanent elements. It was believed that several fundamental principles about the construction of the world were fixed and unchangeable. Although the list of these principles was never completely and explicitly agreed upon, it was commonly implied that such propositions as "the external world exists" or "the future states of the material world are determined by its past states" were constitutive of the scientific mosaic. There was, of course, a serious disagreement between rationalists and empiricists as to how exactly these principles were to be justified. But even in the mid-eighteenth century, when the problematic character of any such principles was made explicit by Hume, it was still commonly accepted that without such fundamental principles there could be no proper knowledge of the world, no science.

This infallibilist picture was abandoned at least by the early twentieth century, when it became obvious that no theory in empirical science can be apodictically true. But even when *fallibilism* became accepted, even when it was understood that no empirical theory can secure its place in the mosaic once and for all, the idea that science has some fixed characteristic features was not abandoned.[5] Since the times of Whewell and Herschel and up until the 1970s it was accepted that one transhistorical feature of science is its method. Although there was little agreement on what this method was and how exactly it had to be explicated, it was generally agreed that there is such an unchangeable method. The last great philosophical conceptions that presupposed the existence of such an unchangeable method were those of Lakatos and the early Laudan.

Some exceptions notwithstanding[6], it is commonly held nowadays that methods are changeable just as theories are. Now, as I have already shown, the absence of an unchangeable scientific method doesn't threaten the prospects of TSC. What seems

---

[4] Galison (2008, p. 111).

[5] A nice account of this transition is provided in Laudan (1981, pp. 111–140); for a shorter version, see Laudan (1996, pp. 211–215) and Lakatos (1970, pp. 10–12).

[6] Elie Zahar is one such exception. See Zahar (1982).

to be vital, however, is the existence of *universal laws* that govern changes in methods and theories. The argument from nothing permanent is dangerous not because it disallows any fixed theories or fixed methods, but precisely because it also assumes that there is no universal mechanism that is responsible for changes in the mosaic. Obviously, if it turned out that there is indeed no such mechanism, a general TSC would become impossible. Thus, it is crucial for the prospects of TSC to consider the second premise of the argument and determine whether *the nothing permanent thesis* holds water.

We can certainly grant that there are neither unchangeable theories nor unchangeable methods. But how have we come to learn (*if* we have) that nothing whatsoever is permanent in science? What exactly is this conclusion based on? Is there anything in our current mosaic of accepted theories that precludes the possibility of anything permanent in science? In order to answer this question we shall find out whether *the nothing permanent thesis* is implicit in any of our accepted theories. *The nothing permanent thesis* would be legitimate if it were indeed found to be a consequence of some parts of our contemporary mosaic. To understand this point, take the case of astrology. The reason why astrology is exiled from the scientific mosaic is that it presupposes a certain kind of influence of celestial phenomena exerted upon physical and mental makeup of human beings – something which is virtually impossible in light of our accepted fundamental theories. Unlike the days of the Aristotelian-medieval mosaic of which astrology was a prominent part, nowadays, if we were to accept any astrological theory into the scientific mosaic, we would introduce incompatibility into the mosaic, since the fundamental premises of astrology are in conflict with the principles of our accepted physical theories, which suggest that even our household electronics have far more significant effect on human body and brain than distant planets and stars. Similarly, if the very existence of the laws of scientific change were incompatible with the theories of our contemporary scientific mosaic (e.g. accepted sociology), the idea of TSC would become extremely dubious. Thus, the crucial question that must be addressed is: do any of the theories accepted nowadays make the existence of the laws of scientific change impossible?

Since *the nothing permanent thesis* has been championed mainly by historians, HSC would be one obvious place to look for a backing of the thesis. We should check, therefore, whether our HSC has shown the inherent changeability of each and every feature of science. Coming across the repetitive claims about transitory character of science, one may get an impression that it is due to historical scholarship that we have come to appreciate that science has no transhistorical features, that it is essentially space and time bound. This impression, I believe, is utterly groundless. In fact HSC cannot possibly show that *no feature whatsoever* is permanent, for that is itself a *general* hypothesis about science. At best, HSC may say that this or that feature *seems to be* present in one historical period and absent in another historical period. It can establish only the *apparent* dissimilarity of two or more historical episodes; this much can be granted. But only a *general* theory of the process, could possibly say anything about the presence or absence of any universal features.

To illustrate this point, we can consider one famous case. Suppose we are faced with two phenomena – a falling apple and a small point of light revolving in the heavens. Suppose also that we have no physical theory whatsoever that would account for these phenomena. Question: how can we tell whether these two phenomena have anything in common or whether they are totally different? Of course, if there were general theories that would explain why apples fall down and why those small celestial light-points revolve the way they do, then it would be possible to say whether they have anything substantial in common or whether they are completely dissimilar. But how can we know whether they have something substantial in common in the absence of any general theory? It is obvious that we cannot; only a general theory can possibly tell us something about substantial similarity or dissimilarity of any two phenomena. Thus, in the Aristotelian natural philosophy the two phenomena were subsumed under two distinct classes. An apple would be an instance of a terrestrial body made predominantly out of heavy elements *earth* and *water* and governed by Aristotelian laws of natural and violent motion. As for the small light-points in the heavens, they were classified as celestial phenomena (either planets or fixed stars) made of the element *aether* and were thought to be moving in circles around the center of the universe. Therefore, if we were to answer the question of their similarity in the 1500s or 1600s, we would definitely say that the two have nothing in common, for they pertain to two different realms – terrestrial and celestial – and are governed by completely different laws. On the contrary, both the Cartesian and Newtonian natural philosophies considered these two phenomena as similar in one important respect. All the differences between the Cartesian and Newtonian theories aside, both posited that falling apples and revolving planets are governed by the same *laws*, for there is no fundamental distinction between celestial and terrestrial bodies. In short, what was completely dissimilar in the Aristotelian natural philosophy became similar in the Cartesian and Newtonian theories. To put it in more general terms, whether there are or aren't any substantial similarities between any two phenomena depends on accepted general theories. Thus, the seeming diversity of historical episodes is not to be taken as sufficient reason for denying the possibility of general laws of scientific change. All the more so because, at first sight, any two given historical episodes seem to have much more in common than the behaviour of a falling apple and a revolving planet. The question of similarities/dissimilarities simply cannot be answered if the mosaic doesn't contain a respective general theory. In the absence of a general theory, we can only point out *apparent* similarities and *apparent* differences. For a deeper analysis, a general theory is needed.

But if HSC doesn't support *the nothing permanent thesis*, where else could it possibly stem from? It may yet turn out that the thesis is implied by our contemporary sociological theories (SOC). Does *the nothing permanent thesis* indeed follow from our contemporary SOC? It may appear at first that there are some sociological grounds for denying the possibility of general laws of scientific change. Consider the following often repeated argument:

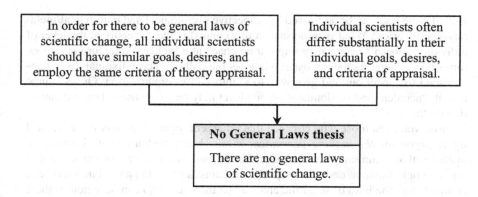

On the one hand, so the argument goes, general laws of scientific change presuppose that scientists working in different fields and at different times share some basic goals and desires. On the other, it is a fact that individual scientists quite often differ from each other in this respect: some pursue academic positions, others want to be renowned, yet others simply work to pay off their mortgages. Therefore, it is argued, there are no grounds to believe that there are any general laws governing scientific change.

Although the argument may appear convincing, it obviously stems from the confusion of two levels of organization – the level of individual scientists and the level of the scientific community.[7] In fact it denies the very existence of the scientific community as an entity with its own emergent properties and emergent patterns of behaviour. It assumes that individual scientists and their individual goals are all that there is. But is this individualistic (atomistic) view in accord with the contemporary SOC? It is safe to say that it isn't.

It is true, of course, that not only scientists but people in general differ in their goals. But these differences do not preclude the possibility of patterns that emerge and function at the social level (i.e. at the level of communities). On the contrary, it is commonly held nowadays that there are community-level regularities, which "emerge along with new social systems, much as chemical laws emerge with new chemical compounds."[8] In fact, if there were no social regularities, no social science dealing with the community as a whole would be possible. Such disciplines as (positive) economics, sociology, or political science would become simply impossible. In addition there could be neither demographic nor tax-revenue forecasts, neither meaningful social engineering nor economic policy.

The same applies to general laws of scientific change: that scientists differ in their personal goals and desires is not a good reason to deny the possibility of higher-level regularities – the regularities that emerge and function at the level of the scientific mosaic. It is not the task of this metatheory to show how exactly these higher-level regularities may emerge from the collaboration of individuals; this is a

---

[7] See section "Individual and Social".

[8] Bunge (1998, p. 33).

factual issue and is to be addressed by respective sociological and ontological/metaphysical theories[9]. What is important in this context is that diversity at the level of individuals doesn't necessarily imply total chaos at a higher level. Our contemporary science suggests the opposite: when people collaborate, individual discrepancies tend to cancel out and give way to a higher-level organization. Or, as Bunge puts it, "accident and randomness on one level may be respectively law and causation on the next"[10].

However, even those who concede that emergent regularities may exist can still try to argue for *the nothing permanent thesis*. Many authors have claimed, for instance, that community-level regularities are necessarily *local*, in the sense that they can only function under very specific circumstances.[11] In particular, it has been claimed that "the laws of scientific change" (if there are any) can only hold if there are certain communities of people, pursuing very specific goals, organized in a very peculiar manner etc. If and only if these and other related conditions are satisfied can there possibly exist laws of scientific change. The conclusion that has been drawn from this is that, strictly speaking, there are no truly general patterns of scientific change, for all regularities are necessarily local.

Although I completely agree with the premise of the argument, I disagree with its conclusion. That the laws of scientific change (provided that there are any) can only function in specific conditions is beyond question. But this is not unique to the laws of scientific change. In fact, all social regularities hold only under very special circumstances, for society itself can exist only in a quite unique physical, chemical, and biological environment. The same goes for the laws of chemistry and biology. The evolution of species (and therefore the laws that govern it) is possible only in a limited range of physical and chemical conditions. The situation with the laws of scientific change is similar: in order for there to be a scientific mosaic of accepted theories there should be a scientific community which can only function under specific social conditions, such as relative stability, supportive educational system, certain degree of autonomy from regime etc.[12] In this sense, the case of the laws of scientific change is no exception.

Thus, if the locality of laws were sufficient reason to void their lawfulness, we would have to give up almost all scientific theories, except those of fundamental physics. For only the laws of fundamental physics are non-local, in the sense of not being bound to specific conditions. The choice is simple: either we have to reject all our theories except quantum physics and general relativity, or we have to admit that the locality of laws is not a vice. Whatever our choice, it is obvious that the laws of scientific change stand or fall together with chemical, biological, and sociological laws.

---

[9] For Bunge's position, see his (1998, pp. 23–26). See also Barseghyan (2009).

[10] Bunge (1998, p. 24).

[11] "Locality" in this sense is not to be confused with "locality" in the sense of being space- or time-bound.

[12] There is an open interdisciplinary question here: what are the social conditions necessary and sufficient for the existence of a scientific community and scientific mosaic?

In brief, I do not see anything in the contemporary mosaic of accepted theories that would support *the nothing permanent thesis*. Neither HSC nor SOC provide any support to the view that there can be no laws of scientific change. Moreover, there are several reasons to think otherwise. Perhaps the most evident indication of this is that the mosaic of scientific theories and methods seems to have a capacity to persist even when cultural, political, social, and economic circumstances drastically shift.[13] Thus, the key elements of the Aristotelian natural philosophy were present in the scientific mosaics of the tenth-century Baghdad, the fourteenth-century Paris, and the seventeenth-century Florence, although nobody will deny the substantial differences among the respective social contexts. Similarly, the current scientific mosaic contains propositions not only from general relativity or quantum physics, but also many elements which have been part of the mosaic in the nineteenth, eighteenth, or even seventeenth centuries. Take, for instance, the second law of thermodynamics, which has been in the mosaic since the nineteenth century. Some elements of the mosaic are even older. Consider, for instance, our belief that "artificial" things (mechanisms, instruments, devices etc.) are subject to the same laws as "natural" things – the belief that was central in both the Cartesian and Newtonian natural philosophies. Or take our conviction that nature can be studied not only by observation but also by experimentation – the conviction that has been an essential part of the mosaic since the rejection of the Aristotelian natural philosophy in the late seventeenth century. When we analyze our current mosaic carefully, we find in it even some ancient "relics" such as the view that the Earth is spherical.[14] It is fair to say that the scientific mosaic appears more persistent than the underlying cultural, social, political, and economic circumstances. It is this apparent persistence of the mosaic that suggests that there could be general patterns of scientific change. Note, that at this stage we can speak only of *possibility*: whether there are *in fact* such patterns is for actual research to establish. However, this is sufficient for our purposes. What is essential is that *the nothing permanent thesis* does not follow from our currently accepted theories. This voids the argument *from nothing permanent*.

The Argument from Social Construction

The idea of constructing a general TSC has been challenged even by those who do not deny the possibility of the laws of scientific change. There is a serious argument against the possibility of TSC, the argument *from social construction*:

---

[13] This point has been made by many authors. See Hacking (1999, p. 87), Weinberg (2003, pp. 135–136).

[14] As I have noted in section "Time, Fields, and Scale", it is not clear whether we can legitimately speak of the scientific community or scientific mosaic prior to the modern period. Therefore, we should be very cautious when speaking about the pre-modern "scientific mosaic".

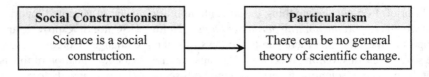

I have chosen this incomplete formulation deliberately, for the argument is rarely presented in a less ambiguous form. The vagueness of the argument stems from the fact that both "science" and "social construction" can be understood differently. In fact, depending on the meanings of "science" and "social construction", we can end up with drastically different lines of reasoning. It is important, therefore, to consider different meanings of both "science" and "social construction".

In the *Introduction* to his *The Scientific Revolution*, Steven Shapin writes: "science is a historically situated and social activity"[15]. Now, if "science" denotes the respective *practice*, then the argument is almost vacuous. For nobody denies that, after all, theories are constructed, published, discussed, and evaluated by social groups who are immersed in their specific cultural, social, political, and economic contexts. It is obvious that the scientific community exists not in a vacuum, but in very specific social circumstances. That is not what is at stake. It is safe to say that, in the argument *from social construction*, "science" denotes not the activity but its *end product* – the scientific mosaic. And it is this meaning of "science" that makes the argument non-trivial. The core idea of the argument is that not only scientific practice, but also the scientific mosaic itself (and all the transitions in it) is in some sense a social construction.[16] We shall therefore modify the argument:

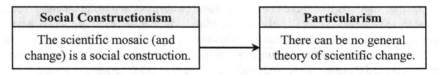

The argument is still rather vague. To make it clear we have to explicate the meaning of "social construction". In what sense is the scientific mosaic said to be a social construction? In his analysis of the term, Ian Hacking suggests that when applied to science the term "social construction" can involve three different aspects – three theses which answer three different questions.[17]

First, the scientific mosaic can be said to be a social construction in the sense that the evolution of the scientific mosaic is contingent, not strictly deterministic. It has been suggested by many constructionist authors that the evolution that has led to the current state of the scientific mosaic is in no way inevitable. They hold that the mosaic could have evolved into something radically different from

---

[15] Shapin (1996, p. 9).

[16] The debating parties often seem to be confusing this important distinction. For discussion, see Hacking (1999, pp. 66–68). Apparently, those who argue that science is not a natural kind refer to scientific *practice*. See Rorty (1988).

[17] Hacking calls them "sticking points". See Hacking (1999, pp. 63–99).

what it is at the moment. We could have had, they say, an accepted physical theory as successful as our electrodynamics which contained no Maxwell equations. Or we could have had an accepted theory of micro-world that mentioned no quarks and even no microparticles at all. On this view, the stages through which the mosaic evolves are not inevitable. For instance, it is possible to conceive of an alternative history where the Newtonian corpuscular theory of light is replaced immediately by some quantum theory of light without intermediate stages of Fresnel's wave theory or Maxwell's electrodynamics. Hacking calls this the *contingency thesis*.[18]

Secondly, the scientific mosaic can be said to be a social construction also in the sense that the theories of the scientific mosaic do not represent the inner structure of the world. It has been claimed by constructionists that our theories cannot possibly represent the structure of the world, for the world is so autonomous and so much concealed that, strictly speaking, it doesn't have any structure. Our theories, on this view, can be quite successful at dealing with the world of experience, but the structures, processes, entities, and properties that they posit are merely our inventions (thus, constructs), for the world doesn't have any inherent structure. Hacking is right to point out that this is the traditional *nominalist thesis*.[19]

Finally, the mosaic can be said to be socially constructed in the sense that changes in the mosaic cannot be fully accounted for without a reference to social factors such as interests, power, networks etc. On this view, scientific change is in some sense reducible to underlying social interactions. In what sense exactly can changes in theories and methods be reducible to underlying social processes? I shall clarify this shortly. At this point, let us agree to label this view the *reducibility thesis*.[20]

Depending on which of these three theses (*contingency*, *nominalist*, *reducibility*) is concealed under the label of "social construction", we end up with three related but not identical arguments. Thus, in order to analyze the argument from social construction we shall consider all the three theses in turn.

Let us start with the *contingency thesis*. What happens with the argument if we understand "social construction" in the sense of the contingency of scientific change? Does it follow from the contingency thesis that there can be no general TSC? We can note that the thesis itself is a factual proposition about scientific change. Namely, it assumes that the course of transitions from one accepted theory to another or from one method to another is not inevitable. It states that given the current state of the mosaic and available contenders, the scientific community could

---

[18] See Hacking (1999, pp. 68–80). The contingency thesis is probably the central message of Pickering (1984).

[19] See Hacking (1999, pp. 81–84). This sense of social construction has been emphasized in Latour and Woolgar (1979/86).

[20] Hacking's own formulation of this point is much weaker. In Hacking's view, all the social constructionist is saying here is that "explanations for the stability of scientific belief involve, at least in part, elements that are external to the professed content of science" (Hacking 1999, p. 92). This formulation with "at least in part" is insufficient, for even Lakatos and Laudan would agree that "elements external to the content of science" should sometimes be employed in the explanation of scientific change. See Laudan (1996, pp. 183–209).

end up accepting different theories; it is conceivable, at least in principle, that we could have had a theory of micro-world without positing any elementary particles or waves, or a biological theory without principles of evolution and natural selection. It is readily seen that not only is this factual proposition in perfect accord with the possibility of TSC, but it is itself a general descriptive proposition about scientific change. Moreover, we know that many authors who were involved in constructing general TSC-s, subscribed to the contingency thesis. Among the proponents of this thesis is Thomas Kuhn.[21] Laudan also stresses this point in his later theory. He says explicitly that neither factual nor methodological disagreements are always conclusively solvable.[22] This amounts to saying that the next state of the mosaic is not strictly determined by its current state, which is what the *contingency thesis* is all about. In short, the contingency thesis is no threat to our project. Quite the contrary, the contingency thesis is itself a general proposition about scientific change, which has already been incorporated in several theories of scientific change.[23]

It is time, therefore, to move on to the *nominalist thesis*. What if we accepted the nominalist thesis, according to which, the structures (objects, properties, processes etc.) posited by our scientific theories are wholly within our representations? Would this affect the prospects of TSC? Suppose for the sake of argument that it would. Suppose that this would somehow make TSC impossible. It is obvious that this would void the possibility of not only TSC but, frankly, any theory about any aspect of the world. If the nominalist thesis were fatal to TSC, it would also be fatal to our physics, chemistry, biology, sociology, and basically any other theory. In other words, the nominalist thesis can in no way be levelled *exclusively* against the possibility of TSC, for if the nominalist thesis were indeed threatening, it would be so not only to TSC, but to other scientific theories as well. Whether the nominalist thesis succeeds in undermining the possibility of our theories is a separate topic that I shall not discuss here. Suffice it that, in this respect, TSC is no different from physical, chemical, biological, or sociological theories – they stand and fall together.

Let us now turn to the *reducibility thesis*. It is, I think, in the sense of the reducibility thesis that "social construction" is most often understood. According to social constructionists, the level of the scientific mosaic and scientific change is in some sense reducible to the level of social interactions, processes, interests etc.[24] But before we can move on to the question of the prospects of a general TSC in light of the reducibility thesis, we have to clarify the very notion of "reduction".

According to the currently accepted view, "reduction" (and, correspondingly, "reductionism") comes in three major varieties – *ontological* (metaphysical), *epistemic* (theoretical), and *methodological* (pragmatic). Therefore, the reducibility thesis (and, correspondingly, "social construction") can acquire three different

---

[21] See Kuhn (1977, pp. 324–329).

[22] See Laudan (1984, pp. 26–45).

[23] It is also one of the theorems of the theory of scientific change presented in *Part II*. See section "Scientific Underdeterminism".

[24] Note that in the social constructionist context, we can speak of reducibility to the *social* level. The reducibility thesis in this limited sense is not to be confused with more general reductionist positions.

meanings, depending on which notion of "reduction" is employed. The three respective theses could be formulated thus:

Ontological Reducibility thesis	Epistemic Reducibility thesis	Methodological Reducibility thesis
The scientific mosaic and scientific change cannot exist independently of the underlying social interactions.	The axioms and theorems of a theory of scientific change can, in principle, be reduced to the laws of sociological theories.	The scientific mosaic and scientific change are most fruitfully studied not by a theory of scientific change, but by sociology.

Obviously, these three theses state quite different things and, therefore, should be treated separately.

According to the *ontological* thesis, the scientific mosaic is a social construct in the sense that all its elements – theories and methods – exist only as long as there exist social groups, processes, and interactions. Namely, the mosaic exists only as a function of the scientific community and not independently of it. This thesis is essentially analogous to other ontological reductionist theses such as "all social systems consist of only human beings and their interactions", "all biological organisms are made of only chemical constituents and their interactions", or "all chemical compounds are constituted by nothing but physical elements and their interactions". The core idea of ontological reductionism in general is that, at higher levels of organization, there is no new type of stuff (no new non-physical forces, no new substances etc.). In its current version – called *ontological physicalism* – it assumes that all that exists is eventually made of only physical stuff or, as they say nowadays, everything supervenes on the physical. It is safe to say that ontological reductionism (in its physicalist version) is the currently accepted view.[25] Similarly, it would be wrong to deny that the existence of scientific mosaic requires the existence of the scientific community which, after all, consists of individual scientists and their interactions.[26] It is perfectly clear that, ontologically, the processes of acceptance and rejection of theories cannot take place without underlying social processes and interactions. In order to get accepted, a theory should be constructed, published, discussed, and evaluated – all essentially social processes. This much is quite straightforward: nobody doubts that, from the ontological standpoint, scientific theories and methods are social products.[27] In this *ontological* sense, the scientific mosaic *is* obviously a social construction.[28]

---

[25] See Stoljar (2009) and references therein.

[26] Note that, in the social constructionist context, the ontological reductionist thesis doesn't go further. It doesn't claim that social processes are, in essence, psychological, biological, chemical or physical. The thesis only refers to the relation of two levels – the level of the *scientific mosaic* and the level of *social interactions*.

[27] See Hacking (1999, p. 67).

[28] Whether it is merely a social construction or whether it has some genuinely emergent properties is another issue, the solution of which doesn't affect our discussion.

But does this mean that the whole idea of TSC is misguided? Does TSC become implausible when we subscribe to the *ontological reducibility thesis*? The answer is an unequivocal "no". The fact that some higher-level systems are essentially made of lower-level elements doesn't imply that there can be no higher-level theory of this higher level of organization. Our contemporary science provides a significant number of examples. For instance, it is currently accepted that, from the ontological standpoint, biological systems are made of only interacting molecules. And yet we do not take it as grounds for abandoning our biological theories. Likewise, the fact that all chemical structures are made of interacting particles doesn't make our chemistry redundant. Therefore, the *ontological reducibility thesis* doesn't endanger TSC – the fact that the scientific mosaic is, in the ontological sense, a social construct doesn't affect the prospects of TSC.

Let us turn therefore to the *epistemic reducibility thesis*. The reducibility thesis in its epistemic version states that even if we succeeded in formulating a theory of scientific change, its laws (axioms, theorems) would necessarily be reducible to more fundamental sociological laws[29]. Say we devised a law of scientific change that states "scientific theories become accepted only if they satisfy the requirements of the currently employed method". Now, according to the *epistemic reducibility thesis* this law is, in principle, reducible to the underlying sociological laws (whatever the latter may be). The same, according to the *epistemic reducibility thesis*, goes for any conceivable law of scientific change. This thesis is analogous to other epistemic reductionist theses such as "the laws of chemistry are reducible to the laws of fundamental physics", "the laws of biological evolution are reducible to the chemical laws", or "the laws of psychology are reducible to the laws of physiology and, consequently, to those of chemistry and physics".

It should be noted from the outset that the possibility of *epistemic* reduction doesn't follow from *ontological reductionism*. Logically speaking, epistemic reductionism presupposes ontological reductionism (physicalism), but not vice versa. Thus, it is conceivable to subscribe to the ontological thesis and reject the epistemic thesis. For instance, there is no internal contradiction in saying that objectively all living organisms are made of only physical stuff and claim that our biological theory is nevertheless irreducible to contemporary physics. Indeed, whether our biological theories are reducible to those of chemistry, whether psychology is reducible to physiology, or whether sociology is reducible to economics or psychology is, nowadays, highly debatable. This is in part due to the vagueness of the notion of "epistemic reduction" – we do not have an accepted view on what it means to reduce one theory to another. Thus, unlike the conception of ontological reductionism, that of epistemic reductionism is far from accepted.[30] Therefore, at this moment it is quite impossible to say what the argument *from social construction* could possibly imply, if "social construction" were understood in the sense of TSC being reducible to sociology (SOC). Indeed, if the social constructionist premise is understood as

---

[29] Whether we will ever succeed in formulating such laws is another issue. For discussion, see Bunge (1998, pp. 222–230).

[30] See Brigandt and Love (2008) and references therein.

saying that TSC is reducible to SOC, then the whole argument becomes extremely vague, for it isn't clear what this "reduction" could possibly mean. In short, the *epistemic reducibility thesis* is not precise enough to do any serious damage.

We may, however, proceed hypothetically as if there were a clear-cut notion of "epistemic reduction". Once we agree on the notion of "epistemic reduction", we may in fact find that the laws of TSC are ultimately reducible to those of SOC (provided that there exist both the laws of TSC and SOC). Such a scenario is conceivable, just as it is conceivable that sociological laws themselves may turn out to be reducible to those of, say, economics or psychology, and then to biology, chemistry, and ultimately fundamental physics. The possibility of this epistemic reductionist scenario cannot be denied on *a priori* grounds. But just as the possibility of this reductionist scenario does not preclude us from searching for specific biological, psychological, economic, or sociological laws, so it should not preclude us from constructing a TSC. The two tasks should not be confused: searching for the laws of scientific change and constructing a TSC is one thing, attempting to reduce these laws to those of SOC is quite another thing. In short, the *epistemic reducibility thesis* should not be taken as an obstacle for TSC.

It is the *methodological reducibility thesis* that is arguably the most threatening of the three. It comes in two versions – strong and weak. The *strong* version basically says that the process of scientific change is fruitfully investigated *only* at the level of social interactions, interests, conflicts etc. On this view, it is SOC and not TSC that can fruitfully study changes in the scientific mosaic. This thesis is analogous to other theses of *strong methodological reductionism*, such as "biological processes are studied fruitfully *only* at the level of interacting molecules", "psychological phenomena are fruitfully studied *only* at the level of neurons and synapses", or "social processes are fruitfully studied *only* at the level of individuals and their interactions". As for the *weak* version of the thesis, it says virtually the same, except that it omits the word "only". According to the weak version of the methodological reducibility thesis, the process of scientific change is fruitfully investigated at the level of social interactions, interests, conflicts etc. The key difference between the strong and weak versions is that the weak version doesn't deny the fruitfulness of non-reductionist strategies. While the strong thesis nowadays is highly debatable, the weak thesis is considered almost a truism.

It is obvious that the methodological reducibility thesis in its weak version can in no way endanger the possibility of TSC. To accept that the sociological study of scientific change is fruitful does not imply an impossibility of a higher-level theory of scientific change. Of course, it would be inappropriate to deny the fruitfulness of reductionist strategies. It is a historical fact that, so far, the strategy of investigating the lower-level mechanisms of higher-level processes has been very successful. Recall, for instance, the success of biochemistry, genetics, or molecular biology. The crucial point here is that the weak thesis doesn't undermine the possibility of theories that study higher-level features. The success of molecular biology doesn't make higher-level theories, such as zoology or ecology, pointless.

It is not the weak but the strong version of the thesis that is threatening to our project. Accepting that changes in the mosaic can be studied *only* by sociology would amount to admitting that the project of TSC is pointless. But should we accept that the reductionist strategy is the only fruitful approach? Is the strong thesis inevitable? I believe that it is not. Consider that the strong thesis represents a danger precisely because it assumes that any higher-level approach is pointless. Thus, if we were to accept the strong thesis, we would have to abandon not only the idea of TSC, but pretty much all our theories except those of fundamental physics. In reality, however, we do not rush to discard our higher-level theories. And this is a perfect indication that we do not presently subscribe to the strong thesis. In other words, it is safe to say that the strong version of the methodological reducibility thesis is not currently accepted. This means that there is no reason to claim that the sociological study of scientific change is the only fruitful approach. For just as the existence of biochemistry doesn't make evolutionary biology or ecology pointless, so the existence of the sociology of scientific knowledge cannot endanger the prospects of TSC.

Thus, we may conclude that none of the three theses – ontological, epistemic, or methodological – undermines the possibility of TSC. The *ontological* thesis is safe, for it only states that all higher-level systems are made of lower-level stuff – it doesn't endanger the status of higher-level theories, i.e. it doesn't assume that we have to throw away all our theories except those of fundamental physics. As for the *epistemic* thesis, it is extremely vague for we still lack an accepted notion of "epistemic reduction". Finally, the *methodological* thesis is nowadays accepted only in its weak version, which in no way endangers the status of higher-level theories such as TSC. Consequently, this voids the argument *from social construction*.

The Argument from Bad Track Record

The final argument against the possibility of TSC that I shall consider is the argument *from bad track record*. Nowadays it is taken for granted that there have been too many unsuccessful attempts of constructing TSC. It is sometimes argued that, since we have repeatedly failed to construct anything worthy, further attempts are simply pointless:

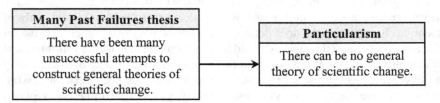

One can point to the problematic character of the kind of reasoning from unsuccessful past attempts to the inexpediency of any future attempt. This conclusion, I believe, is flawed, for HSC gives us many illustrations of the opposite view. Normally, initial failures are not taken as good reason for despair. Physicists, for instance, do not stop searching for general laws when their initial attempts fail to produce the desired results. The same is true, for example, in computer science where "pessimism based on the failure of few, particularly simple, programming attempts is not taken seriously, and properly so"[31]. Thus, even if we grant that there have been some genuine theories of scientific change, it shouldn't preclude us from trying to construct a new TSC.

The argument has several flaws. For instance, it can be noted that past failures do not logically imply future failures and that the whole implication is unsound. The main flaw of the argument, however, lies in its premise. If we stick to the notion of TSC developed in chapter *Scope* above, it becomes obvious that we have barely had any theory that could count as a genuine TSC. For one thing, most of the theories that have attempted to explain scientific change date back to times when the method of science was thought to be unchangeable. As a result, most of these theories stem from the confusion between TSC and MTD.[32] In addition, most of the extant theories fail to differentiate the social and individual levels.[33] Moreover, many of these theories mix up *acceptance*, *use*, and *pursuit*.[34] In this sense, arguably the closest thing to a general TSC to date is the later Laudan's *reticulated model*. And even this reticulated model, strictly speaking, cannot count as a genuine TSC, since it doesn't explain changes from one employed method to another, but only from one openly prescribed methodology to another.[35] So it wouldn't be an exaggeration to say that we are yet to construct our first genuine TSC, a general theory that would explain the mechanism of transitions in theories and methods alike. Therefore, to say that we have exhausted all the possibilities would be highly misleading. This voids the argument *from bad track record*.

Hitherto I have deflected several arguments against the possibility of TSC. It is clear that the existing particularist arguments do not endanger the project of constructing a general TSC. Of course, at this stage, it would be premature to argue that eventually we will have such a theory – our efforts may yet turn out fruitless. Yet, the good news is that there seem to be enough similarity among different historical episodes to suggest that the process of scientific change might be governed by a certain set of laws.[36] Whether there are actually any general laws of scientific change is for actual research to establish; at this point we must appreciate that there is no reason to think that TSC is doomed to failure.

---

[31] Kelly (2000, p. 179).

[32] Take the early Laudan's theory, for example, which implies the existence of the static method of science and, consequently, says nothing about changes in methods. See section "Descriptive and Normative".

[33] Recall the examples discussed in section "Individual and Social".

[34] Insofar as I know, no extant theory distinguishes between the three. See section "Acceptance, Use, and Pursuit".

[35] See section "Explicit and Implicit".

[36] Another reason for optimism is that actual scientific activities seem to be more restricted than those in economics where we have fairly successful theories. See Giere (1984, pp. 27–28).

Chapter 3
Assessment

No doubt, it is a relief to know that our project is possible or, rather, that it's not impossible, i.e. that it's not doomed to failure. Now, suppose we have constructed a TSC – a general descriptive theory of changes in the mosaic of accepted theories and employed methods. Question: how should we *assess* that TSC? There are two sides to this question. On the one hand, it is a question of *method*: what conditions should the TSC satisfy in order to get accepted, i.e. what is the method of assessment of the TSC? On the other hand, it is a question of *relevant* data: what kinds of facts are relevant to assessing that TSC? In this section, I shall address both sides of the issue.

Method

As I have indicated earlier, when it comes to the method of appraisal in a certain field of inquiry, we can always ask two distinct questions – one descriptive and one normative:

Descriptive question	Normative question
? What method *is* currently employed in theory assessment?	**?** What method *ought to* be employed in theory assessment?

The former is a descriptive question concerning the actual method currently employed in theory assessment. Thus it pertains to the domain of HSC. The latter is a normative question that concerns the *legitimacy* of the method, i.e. it pertains to the domain of methodology (MTD) for it asks what method *ought to* be employed in theory assessment. It is obvious that these two questions are not identical. Of course, it is quite possible for these two questions to have the same answer, but that

© Springer International Publishing Switzerland 2015
H. Barseghyan, *The Laws of Scientific Change*, DOI 10.1007/978-3-319-17596-6_3

doesn't make them indistinguishable. It may turn out that the currently employed method is so flawless that we all agree that it is strongly advisable to continue employing this same method in the future. But the current method may also turn out to have serious drawbacks – so serious that we would refrain from prescribing it. It is important, therefore, not to mix up the descriptive and normative questions.

Naturally, we can ask these two questions with regard to TSC. The descriptive question could be formulated as follows: what method *is* actually employed in assessing a TSC? Namely, what criteria must a new TSC satisfy in order to get accepted? This is the same as to ask: what implicit expectations do we have regarding a new TSC? Correspondingly, the normative question would be: what method *ought to* be employed in assessing a TSC? It is these two questions that I shall discuss in this section. But before addressing these questions, I shall clear up a possible confusion.

It isn't difficult to confuse these two questions of method with analogous questions regarding methodology (MTD). Naturally, one can formulate both the normative and descriptive questions also with regard to MTD. The descriptive question would be: what method *is* actually employed in assessing a MTD, i.e. what criteria must a MTD satisfy in order to become openly prescribed by the community? The normative question would be: what method *ought to* be employed in assessing a MTD? The following diagram illustrates this situation:

	Regarding TSC	Regarding MTD
Descriptive	**?** What method *is* actually employed in assessing TSC?	**?** What method *is* actually employed in assessing MTD?
Normative	**?** What method *ought to* be employed in assessing TSC?	**?** What method *ought to* be employed in assessing MTD?

The normative question regarding MTD has traditionally been ascribed to the domain of the so-called *metamethodology*. It is considered the task of metamethodology to come up with a set of conditions that an acceptable methodology ought to meet. This question was discussed in detail by both Lakatos and Laudan. Lakatos's self-referential approach, where a MTD is supposed to be assessed by its own standards elevated to the level of metamethodology, is one instance of metamethodology. Laudan's normative naturalism with its insistence that a MTD ought to be tested against the historical record is another example of metamethodology. Similarly, by deciding to employ the hypothetico-deductive method for testing methodological theses, the members of the VPI project have thus subscribed to a particular (hypothetico-deductivist) metamethodology.[1]

As for the descriptive question regarding MTD, it is empirical in nature and thus to be tackled by HSC; only a careful study of our current expectations can tell us what criteria we tacitly employ in assessing different methodologies. However, it would be fair to say that this descriptive question has hardly ever been addressed separately from its normative sibling. This shouldn't be surprising, since, as I have explained in section "Descriptive and Normative", those who hold that the core method of science is unchangeable naturally end up confusing normative and descriptive issues.

This confusion has many different manifestations. One such manifestation is the belief that TSC and MTD are two sides of the same descriptive-normative enterprise. It is a traditional conviction that uncovering the mechanism of scientific change amounts to explicating the method of science which both *is* and *ought to* be employed in theory assessment. Another expression of this confusion is a failure to distinguish the descriptive historical task of explicating the implicit criteria of acceptance of MTD from the normative metamethodological task of formulating the list of conditions that an acceptable MTD ought to meet. There is, finally, the third manifestation of this confusion which stems from the first one. If one welds TSC and MTD together, if one holds that uncovering the laws of scientific change is the same as prescribing criteria for theory assessment, then one is destined to spread this confusion to the level of respective methods and hold that TSC and MTD have one and the same method of assessment. Indeed, if TSC and MTD were indistinguishable, then explicating/prescribing the method of assessment of TSC would be the same as explicating/prescribing the method of assessment of MTD. The outcome of this threefold confusion (common to Lakatos, Laudan and many others) is that it produces not four different questions, but one amalgamated question:

Regarding TSC-cum-MTD
Descriptive-Normative **?** What method *is* and *ought to be* employed in assessing a new TSC-cum-MTD ("theory of rationality", "theory of scientific method", "methodology", "philosophy of science" etc.)?

---

[1] For Lakatos's metamethodology see his (1971). Laudan's metamethodology is presented in his (1996), pp. 125–179. For the VPI-project's choice of metamethodology, see Donovan et al. (eds.) (1992), pp. xi–xv, 3–14. For discussion of the major metamethodologies, see Nola and Sankey (2007), pp. 80–103, 252–336.

When Nola and Sankey discuss the question of metamethodology, they seem to have in mind this four-in-one question – the fused normative-descriptive question regarding the method of assessment of the fused TSC-cum-MTD.[2]

The mistake is obvious, for there are not one but four different questions here. The same line of reasoning that makes the difference between descriptive TSC and normative MTD evident also implies that the questions regarding the respective methods of assessment of TSC and MTD are different. To aim at explicating our existing implicit criteria concerning TSC is one thing, to prescribe such criteria is quite another. Similarly, it is one thing to try to explicate the implicit criteria for a good MTD, and it is another thing to prescribe what criteria an acceptable MTD ought to meet. Finally, the current method of assessment of TSC is not necessarily the same as the current method of assessment of MTD (just as the current method of, say, physics isn't necessarily the same as the current methods of psychology, sociology, or cultural studies). Therefore, explicating or prescribing the one doesn't necessarily amount to explicating or prescribing the other. Albeit quite interesting, the questions regarding the method of MTD do not concern us here. Our task is to tackle the two questions concerning the method of assessment of TSC: what is the current method of assessment of TSC and what ought it to be?

In order to answer these questions, it is vital to recognize that TSC is of a twofold nature. On the one hand, it is a theory of a specific process in time, i.e. the process of scientific change. From this perspective, TSC pertains to the theory-level, while its object – the process of changes in the scientific mosaic – pertains to the object-level. It is this perspective that we have been discussing so far:

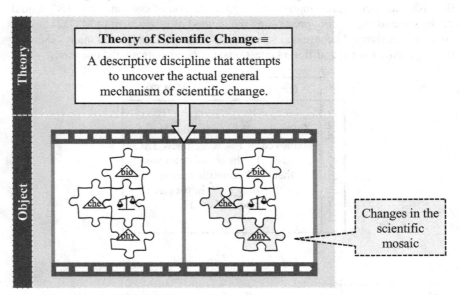

---

[2] See Nola and Sankey (2007), pp. 80–103.

On the other hand, if any particular TSC became accepted by the scientific community, it would itself become part of the scientific mosaic, just as any other accepted theory. From this perspective, TSC is no different from any other theory about any other process in time. An introduction of a TSC, just as an introduction of other theories, is nothing but a suggestion to modify the mosaic by adding to it a new theory. If the proposed modification becomes accepted, the TSC itself becomes part of the scientific mosaic and, thus, belongs to the object-level:

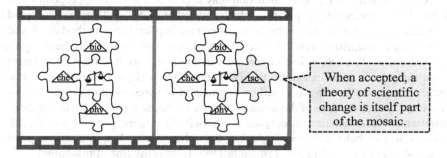

When accepted, a theory of scientific change is itself part of the mosaic.

In short, TSC pertains to both *theory-* and *object-*levels. There is nothing extraordinary about this. In fact, the theory-object relation always depends on a perspective. Something can be an object of study only relative to a certain theory, and vice versa. So, from the perspective of HSC, an accepted TSC is just another element of the mosaic; it pertains to the object-level. But from the perspective of TSC, it pertains to both levels. Thus, TSC is necessarily self-reflective: among other things, any TSC must also explain transitions from one accepted TSC to the next:

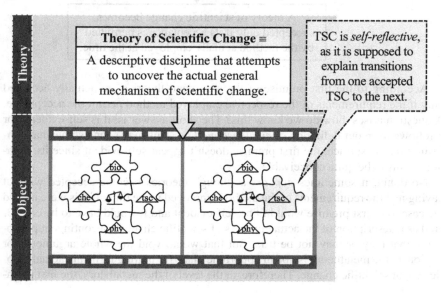

Theory of Scientific Change ≡

A descriptive discipline that attempts to uncover the actual general mechanism of scientific change.

TSC is *self-reflective*, as it is supposed to explain transitions from one accepted TSC to the next.

Once we realize this, we move one step closer to answering the descriptive question. It becomes clear that a TSC will actually be assessed just as any other scientific theory, i.e. a TSC will be expected to conform to the respective requirements of the method employed at the time of the assessment. There is only one way for a TSC to become accepted – it will have to meet the implicit expectations of the community concerning TSC-s (whatever those expectations will be at the time of the assessment).[3]

Although at first sight this conclusion may appear as a normative proposition, it is not. It is a descriptive proposition which states that a TSC will become accepted only if its acceptance is allowed by the requirements implicit in the method employed at the time of the assessment. This descriptive proposition stems from the recognition that, after all, a TSC is itself a scientific theory and, thus, its case is no different from that of any other scientific theory. Thus, when a physicist proposes a new theory, the theory is being assessed by the method of assessment of physical theories employed at the time of the assessment: it becomes accepted if it satisfies those requirements and remains unaccepted if otherwise. The same goes for a chemical, biological, psychological, sociological or any other theory. This is a crucial point which should be treated in a piecemeal fashion. Here is my line of reasoning:

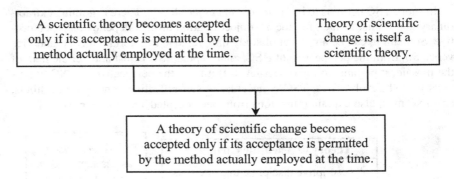

According to the first premise of the argument, theories are actually accepted only if the requirements of the respective employed method permit the acceptance. A question arises: how do we know this? The short answer is: it is self-evident, for it follows from our definition of *employed method*. This may sound somewhat confusing: on the surface, the first premise doesn't appear self-evident since its negation seems to be quite conceivable.

No doubt, it sometimes *appears* as though theories become accepted without having met the requirements of the method employed at the time. If this were indeed the case, our first premise wouldn't be self-evident and it would have to be considered as a description of the actual process of scientific change – a contingent proposition that may or may not be true. But that would void the whole argument, for obviously the metatheory is not an opportune place for putting forth any actual theories about scientific change. Therefore, at the level of the metatheory, the first prem-

---

[3] Some authors seem to understand this. See Hull (1979), p. 11; Freedman (2009), pp. 313, 315.

ise could be conceded only if it were a necessary truth, a tautology that followed from our basic definitions. The key question that would help us determine whether the premise in question is necessary or contingent is: can a theory become accepted without having satisfied the requirements of the method employed at the time? If such a scenario were conceivable, even in principle, the first premise would be a contingent proposition. Conversely, if this scenario turned out to be impossible in principle, then we would say that we deal with a necessary proposition, something that is true simply by virtue of the respective definitions.

Although it sometimes *appears* as though theories become accepted in violation of the current method, in fact the currently employed method cannot be violated. It cannot be violated simply by virtue of the definition of *employed method*. Recall that *employed method* has been defined as a set of conditions that should be met by a proposed theory in order to become accepted. Consequently, a method is said to be *employed* in theory assessment when theories get accepted only if they satisfy the requirements of the method. This is what the notion of "currently employed method" is all about: meet its requirements and become accepted or else remain unaccepted. Therefore, the appearance that a theory has managed to become accepted without satisfying some requirement *r*, can only mean that requirement *r* was not part of the method employed at the time.

What is often violated is not the *method*, but this or that openly expressed set of rules, i.e. one or another *methodology*. Feyerabend is right only in this sense: the actual scientific practice often violates the dicta of our openly prescribed methodologies. This is too obvious to be denied.[4] But to say that the currently employed method was violated is to say something self-contradictory: the employed method cannot be violated simply because it is defined as a set of conditions that should be met by a theory in order to get accepted. Thus, if a theory becomes accepted by seemingly violating our current requirements, it is a good sign that we do not quite understand our own implicit requirements.

Let us take an example. Suppose, we openly prescribe a methodology which stipulates that a theory can be accepted only if it has some confirmed novel predictions. Suppose also that there is a theory that has actually managed to get accepted without any confirmed novel predictions at hand. What should our conclusion be? Feyerabend would readily pronounce that our method has been violated. But this conclusion is incorrect, for it is not the employed method that has been violated, but only the principles of our explicitly prescribed methodology. The only legitimate conclusion to be drawn is that our actual expectations were not what we thought they were – in this case, we weren't actually expecting a new theory to have confirmed novel predictions.

In short, our first premise is necessarily true, since it follows from our definition of *employed method*. Consequently, we have to admit that the whole argument is valid, albeit somewhat trivial. It doesn't say much, of course, for it simply clarifies what was implicit in our definitions. It merely makes explicit that a TSC will become accepted only if its acceptance is allowed by the current method:

---

[4] I have explained this in section "Explicit and Implicit".

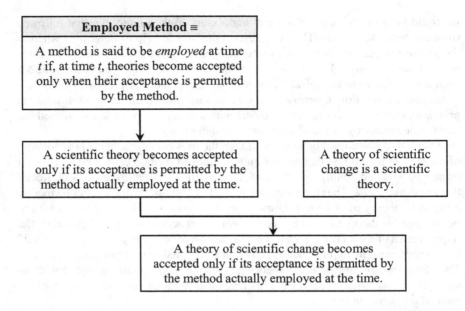

Now, this may be well and good, but an important factual question remains to be answered: what is this currently employed method of evaluating TSC-s after all? How is it to be spelled out? What are the implicit expectations of the community concerning TSC-s? For instance, would it suffice if a TSC provided explanations for all past episodes? Or would it be expected to predict some new cases of scientific change in order to get accepted? In short, what do we actually expect from a TSC?

As I have mentioned earlier, this factual question pertains to the domain of HSC. It is therefore to HSC that we have to turn for the correct explication of our current method. The historical record shows, however, that our attempts to explicate our employed methods, to put it mildly, haven't always been successful. Take, for instance, the eighteenth-century conviction that the actual method of science was correctly summed up in Newton's famous rules. Given the many well-known obstacles to accurate explication, one may legitimately doubt the possibility of explicating our current method with necessary precision. But if we fail to explicate our current method, then how are we to go about assessing our TSC-s? What criteria are we going to employ in evaluating a TSC once it is constructed?

This isn't, however, a serious impediment for our project, for actually such an explication is not necessary. To acknowledge this point, consider the situation in other fields of inquiry. Take the current method of physics, for instance. How can it be explicated? Does it presuppose that only theories with confirmed novel predictions can be accepted? Or does it allow a theory to get accepted even in the absence of novel predictions? And if novel predictions are a must, then how is novelty to be understood – as temporally novel, as heuristically novel, or as something else?[5]

---

[5] These days, there is no agreement as to whether novel predictions are a must and, if so, what exactly the notion of *novelty* amounts to. See Zahar (1973); Musgrave (1974); Lakatos and Zahar (1976); Gardner (1982); Schlesinger (1987); Brush (1994); Leplin (1997); Worrall and Scerri (2001); Hudson (2007) and references therein.

Besides, does the current method imply that, in order to get accepted, a theory should somehow account for all the phenomena accounted for by the currently accepted theory? Also, is it mandatory for a new theory to solve all the problems solved by its predecessor? Explicating the current method of physics amounts to answering these and other similar questions. However, HSC seems to suggest that nowadays there is no unanimity as to how exactly these questions are to be settled. In other words, presently there is no accepted explication of the method of physics. But does this preclude physical theories from getting accepted or rejected by the scientific community? In fact, not at all: the process of scientific change doesn't stop simply because there is no accepted explication of the currently employed method. This isn't surprising, given that it is not our explications but methods themselves that are actually employed in theory assessment.[6] Similarly, in our case, an explication of our tacit expectations is not compulsory, since a TSC can become accepted even in the absence of such an explication. In other words, even if the historian fails to explicate our current method correctly, it will in no way endanger our project.

In brief, the only answer to the descriptive question that the metatheory can give is the one that follows from the definition of *employed method*: a TSC can become accepted only if it satisfies the requirements of the current method. In order to be able to assess a TSC, it is not vital to know explicitly what this current method is.

Let us now turn to the normative question: what method ought to be employed in assessing a TSC? This normative question pertains to the domain of methodology (MTD). Answering this methodological question amounts to prescribing a set of criteria for assessing a TSC. While some of these criteria can be deduced from the very definition of TSC, others can only result from a serious methodological study. The requirements that follow from the definition of TSC are trivial. Just as any theory of free fall is expected, by definition, to explain the phenomenon of free fall, so any theory of scientific change is expected to account for, well, the process of scientific change. Namely, it is expected to account for transitions from one accepted theory to another and from one employed method to another. This much is clear. It is, however, the non-trivial requirements that are most interesting. In particular, should a TSC satisfy the requirements of the Popperian, Lakatosian, Laudanian, Bayesian or some other methodology? Which methodology ought to guide the assessment of a TSC?

It is obvious that in order to answer this question we must adjudicate between several competing methodological conceptions.[7] This, in turn, can lead us to another question: how are we to choose the proper methodology? That is, what metamethodological requirements should a respectable methodology satisfy? It is important to keep in mind that unlike the mid-eighteenth century, or the "good old days" of logical positivism nowadays we seem to have no universally prescribed methodology at all. There are many methodologies available on the market: Popper's methodology of conjectures and refutations, different modification of Lakatos's methodology, the

---

[6] Refer to section "Explicit and Implicit" for discussion.
[7] For discussion of major methodologies, see Nola and Sankey (2007).

early Laudan's problem-oriented methodology, Bayesianist methodologies, etc. In addition, we don't have even a universal metamethodological standpoint, from which we could adjudicate our methodological theories. There are many competing metamethodologies: reflective equilibrium, normative naturalism, conventionalism, etc. In these circumstances, how could we possibly tackle our normative issue? If we indeed were to search for an answer, we would first have to choose among several metamethodological theories, then descend to the level of methodologies and determine the best one according to the metamethodological standards. But such a voyage into the thickets of methodology and metamethodology may certainly lead our discussion astray, for the topic is immense. Many have gone that way, but none have returned.

Luckily, however, settling the normative question is not vital for our project – at least, not at this stage. Indeed, what would change if we had agreed on the methodology of assessment? Suppose for the sake of argument that the community openly prescribed some normative methodology, i.e. a set of openly formulated requirements for assessing TSC-s. Question: would this be enough to affect the actual procedure of assessment? In order to affect the actual procedure of assessment this openly prescribed methodology would have to be capable of changing the currently employed method since, by definition, it is the current method that is being employed in theory assessment. That is to say, our decision to openly subscribe to a set of methodological rules would have to affect our actual implicit expectations concerning TSC-s. Therefore, the question becomes: can our choice of methodology affect our currently employed method? Or, in other words, is it possible to deliberately alter our implicit expectations? Although methodologists usually act as if employed methods were actually shaped by openly prescribed methodologies, the situation is not that straightforward. One can recall many historical episodes, where the openly prescribed methodology had little in common with the then-employed method.[8] At this point we have to acknowledge the following: whether it is or isn't possible to change the currently employed method by prescribing a new methodology is a question that only an actual TSC can address in collaboration with HSC. The answer shouldn't be speculated: we have to postpone answering it until we have a full-fledged TSC.[9] We can conclude, therefore, that at this level – the level of metatheory – there isn't much point in discussing the normative methodological question, for even if we were able to come up with an answer, we wouldn't be in a position to tell whether our preferred methodology would be capable of affecting our actual expectations. The argument can be summed up thus:

---

[8] See section "Explicit and Implicit" for examples.

[9] The actual TSC, proposed in *Part II*, contains a respective theorem. See section "The Role of Methodology" below.

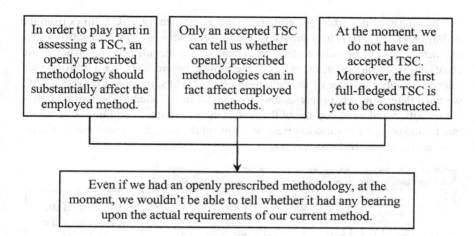

| In order to play part in assessing a TSC, an openly prescribed methodology should substantially affect the employed method. | Only an accepted TSC can tell us whether openly prescribed methodologies can in fact affect employed methods. | At the moment, we do not have an accepted TSC. Moreover, the first full-fledged TSC is yet to be constructed. |

Even if we had an openly prescribed methodology, at the moment, we wouldn't be able to tell whether it had any bearing upon the actual requirements of our current method.

This allows us to refrain from delving into methodological debates concerning the best methodology and, thus, to avoid even more perplexing conundrums of metamethodology. The situation is analogous to that of any other field of inquiry. For instance, we may argue about the method of assessment proper for biological theories. We may quarrel, for example, whether biological theories ought to provide novel predictions. This methodological activity may be quite illuminating and even productive – it is possible that we may end up concluding that one particular methodology is preferable to others. But the crucial point is that in reality our biological theories are being accepted or rejected regardless of whether or not there is any methodology openly prescribed by the community.

This brings us back to the conclusion that we reached at the end of our discussion of the descriptive question: when a TSC is constructed, it will inevitably be assessed by the method employed at the time of the assessment (whatever that method may turn out to be). As for the normative question, the metatheory lacks proper means for answering it. At this point, we know neither how to choose the best methodology nor how our choice can possibly affect our actual implicit expectations. Therefore, we must postpone the discussion of the normative question until after we acquire an accepted TSC.

Relevant Facts

Generally, the assessment of a scientific theory requires not only some method of assessment but also some evidence. Suppose we had a TSC ready to be assessed. What classes of facts would we take into consideration when assessing the TSC? In other words, what phenomena would be *relevant* to the assessment? Normally, the full list of the relevant facts has two determinants. While some phenomena are relevant to the assessment of a theory because the theory itself suggests their relevance, other phenomena are relevant simply by definition.

The relevance of certain phenomena to the assessment of TSC follows from the definition of TSC. To clarify the list of these phenomena, we don't have to wait for any actual TSC to be constructed; the means of the metatheory are sufficient for clarifying it. In particular, the states of the mosaic before and after a transition are relevant simply by virtue of the definition of TSC. Thus, in order to assess a TSC, we must determine what theories and methods were included in the mosaic before the transition and what became of them after the transition. In addition, it is important to find out what contenders were available at the time, i.e. what other modifications to the mosaic had been proposed.

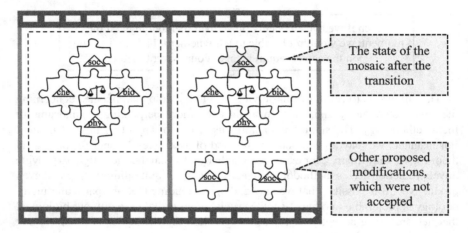

In short, when assessing a TSC we must seek to answer a number of factual questions. What *theories* were accepted before the transition? What *methods* were employed by the time of the transition? What *modifications* were proposed and what parts of the mosaic did they intend to replace? What modifications became *accepted*, i.e. what did the state of the mosaic become after the transition?

However, the relevance of other phenomena can be established only by an actual TSC under scrutiny. Any actual TSC may suggest its own list of phenomena that presumably play a role in scientific change. For example, if a TSC under scrutiny were to posit a connection between the process of scientific change and specific cultural, social, political, or economic factors, then it would be crucial for the assessment of a TSC to uncover what these cultural, social, political, or economic factors were at the time of each transition in the scientific mosaic. The same applies to any scientific theory. If a biological theory under scrutiny declares that the existing diversity of species is produced by an ongoing evolutionary process, then the search for the fossils of extinct species becomes relevant to testing this theory. Similarly, if a physical theory posits the existence of a certain type of microparticle, the detection of that particle becomes relevant to testing that theory. In short, the relevance of some phenomena can be suggested only by an actual TSC.

It is worth repeating that the metatheory can go only this far. It cannot provide us with a *complete* list of all relevant phenomena; for the full list, we shall wait until an actual TSC is constructed. What the metatheory does establish is that there are phenomena which are relevant to the assessment of TSC simply by the definition of TSC. In particular, it tells us that the respective states of the mosaic before and after a transition as well as all proposed modifications are among the relevant facts. This is similar to saying that falling apples are necessarily relevant to any theory of free fall, for that is what follows from the very definition of theories of free fall. But the definition alone cannot tell us what other factors are relevant to the process of free fall – that can be suggested only by an actual theory of free fall. This much is quite straightforward.

As soon as we realize that it is changes in the mosaic that TSC is concerned with, it becomes apparent that our existing historical knowledge is fragmentary. There are two reasons for that. On the one hand, as we will see in section "Indicators", reconstructing the state of the mosaic at a given time is not a simple task. On the other hand, since these days HSC is not being guided by a TSC, reconstructing the state of the mosaic of a given community at a given time is not always the first priority of HSC. As a result, for the most part, our historical knowledge of different scientific mosaics is not precise enough. For instance, many historical narratives concerned with theory acceptance talk about acceptance or rejection of *whole* theories, whereas very often what gets accepted or rejected by the community is not a whole theory but only certain *parts* of it – some propositions get accepted while others remain unaccepted. For instance, it is common knowledge that the Newtonian theory was accepted in Britain by the beginning of the eighteenth century. But are we sure that among the accepted propositions were, say, the idea of a force of gravity acting at a distance or Newton's conception of absolute space and time? In more general terms: do we know exactly what parts of Newton's theory were accepted? A precise reconstruction of different scientific mosaics is, therefore, a task of utmost importance – a task that only a theory-guided HSC can accomplish.

When the details of the mosaic are ignored, we often end up discussing such anachronistic topics as "the religious constraints on science in the seventeenth century" or "the interaction of science and religion in the seventeenth century". By doing this, we simply ignore that back then "religious constraints" were nothing but a set of accepted *theological* propositions. In the seventeenth century, these theological propositions were part of the mosaic just as those of natural philosophy. They weren't something external to the mosaic. The proposition that God created the world was as accepted as the proposition that the Earth is spherical; they both were parts of the mosaic. In order to avoid such anachronisms, we must explicate the consecutive states of the mosaic with utmost precision.

It is true that not all historical studies aim at reconstructing different states of the mosaic, for there is clearly more to the history of science than HSC.[10] However, even when we are concerned with the history of theories and methods, we often focus on aspects that are not directly related to the concerns of TSC. Among the most common questions that we normally discuss in our historical studies concerning theories and methods are "when and under what circumstances did the author come up with an idea/theory?", "who exerted an influence upon the author and in what way?", "what role in the development of her ideas did personal,

---

[10] See section "Epistemology, History and Theory of Scientific Change" for discussion.

social, economic, cultural, or political factors play?", "how did the theory develop
and what led the author to elaborate/modify it?", "when and where was the work
in question first published?", "how many editions has it gone through and what
were the important differences between the editions?", etc. It is not at all obvious
how answering these questions can help us in our quest for a general TSC.

Consider the following extracts from *The General History of Astronomy*, one of
the most excellent collections of its kind.[11] "Gilbert's work was used by Galileo to
establishing two points..."[12] "Following his meeting with Isaac Beeckman in 1618
and the famous dream of 10 November 1619, René Descartes... spent nine years
developing a mathematical physics independent of Aristotle's philosophy."[13] "From
Descartes [Newton] took over the principle of inertia..."[14] One can easily supple-
ment this list with many other similar examples. Common to all these fragments is
that they deal with influences which one *individual* scientist exerted upon the ideas
of another. No doubt, tracing the genealogy of an idea or a theory is quite an interesting exercise in its own right, but it hardly says anything about actual changes in
the scientific mosaic.

Take another set of extracts from the same volume. "By the beginning of 1573,
Tycho had decided that the object was indeed nothing less than a new star."[15] "A
study of the comet of 1577 convinced [Tycho] that [the supposedly solid planet-
bearing] orbs did not in fact exist."[16] "When Galileo examined the Moon... he found
that it was... much like the Earth. To Galileo this was clear evidence that Aristotle
was wrong..."[17] Passages of this sort are also innumerable in the contemporary
HSC. What they all have in common is readily seen: they all deal with reasons
which convinced not the community but individual scientists to accept or reject
certain views. In other words, they deal with changes in individual belief systems
and not changes in the mosaic.

Contemporary narratives share yet another common trait. They are often concerned exclusively with the construction and elaboration of a theory and pay
relatively little attention to the respective changes in the scientific mosaic. It
often remains a mystery under what circumstances a particular theory became
accepted by the community. As our current historical research is not guided by
any TSC, the information crucial from the standpoint of TSC is often presented
in between the lines, in the footnotes, or even omitted from historical narratives
altogether. Perhaps the most vivid illustration of this point is how much we focus
on the history of construction and elaboration of Newton's theory and how little
on the process of its acceptance. We all undoubtedly know how Newton's
Principia was written and when it was published. But when did the Newtonian

---

[11] See Taton and Wilson (eds.) (1989).
[12] Pumfrey (1989), p. 45.
[13] Aiton (1989), p. 207.
[14] Wilson (1989b), pp. 234–235.
[15] Thoren (1989), p. 5.
[16] Schofield (1989), p. 33.
[17] Van Helden (1989), p. 83.

theory (and what individual propositions of it, for that matter) become accepted by different communities? It would be interesting to know how many of us could come up with the correct answer.[18]

It is worth repeating that historical narratives concerned with the *individual* rather than *social*, *construction* rather than *appraisal*, or *methodology* rather than *method* are quite interesting in their own right. There is nothing wrong with such historical narratives since, as we know, there is more to the history of science than the history of the scientific mosaic (HSC). But as far as the assessment of TSC is concerned, they are of little use since often they do not provide data relevant to TSC. Once we decide to study changes in the scientific mosaic, such an approach to history becomes lacking.

To conclude this section, it is worth noting that many authors have realized that there should be some criteria for distinguishing between the relevant and irrelevant facts of history.[19] Unfortunately, not many have realized that such criteria shouldn't be picked up arbitrarily. As I have argued, the complete list of relevant phenomena can be provided only by an actual TSC. However, the relevance of some phenomena can be established already at the level of the metatheory. In particular, the metatheory tells us that, when assessing a TSC, we should necessarily focus on the successive states of the mosaic as well as the suggested modifications (available contenders). In order to test a TSC, we must focus on such questions as "what theories constituted the scientific mosaic of the time?", "what contenders were available?", "which of the contenders became accepted?", "how, why, and when did a particular theory become accepted by the community?", "what method was actually employed in the assessment?", "what became of the mosaic after the transition?", etc.

Indicators

To know the types of relevant facts is one thing, to be able to pin them down is quite another. As we have already established, when studying a particular historical episode, it is vital to reconstruct the precise state of the scientific mosaic before and

---

[18] Consider Cohen's treatment of Newtonian mechanics where he focuses almost entirely on the development of the theory, whereas the circumstances of its acceptance are all but ignored. No wonder that with such a negligence of acceptance he ends up giving a wrong year. According to Cohen, the "validation" (another confusing term) of Newton's theory came in 1758 and had to do with the return of Halley's comet. See Cohen (1985), pp. 182–183.

However, it would suffice to look at the records of the French Académie as well as the respective articles of the famous Encyclopédie to confirm that the theory had been accepted in France long before the return of Halley's comet. In France, the acceptance apparently took place circa 1740, after the famous measurements of the shape of the Earth. See Aiton (1958), p. 172; d'Alembert (1751), pp. 80–83. In Britain, it had been accepted even earlier since circa 1700.

[19] Lakatos (1971) is one example. Some contemporary authors too seem to realize this. See, for instance, Pinnick and Gale (2000), pp. 118–119.

after the transition. But how does the historian establish that a theory was accepted during a specific time period? What are the *indicators of theory acceptance* and what are the *indicators of method employment*? Before we proceed, it must be appreciated that we are about to enter the domain of HSC here, for it is a task of the historian of scientific change to locate the indicators of theory acceptance and method employment applicable to different mosaics and different time periods. Therefore, the indicators suggested in this section are all tentative and for illustrative purposes only.

I shall start from the possible *indicators of theory acceptance*. How does the historian know that a certain theory was accepted at time *t*? For one thing, it is surely not the opinions of individual (albeit great) scientists. As I have stressed in section "Individual and Social", it is not difficult to see what would happen if we were to take opinions of individual scientists as a clear indicator of theory acceptance. Take standard historical narratives on the Scientific Revolution, for instance. Very often, they revolve around the figures of great scientists, like Copernicus, Kepler, Galileo, Descartes, Boyle, or Newton. Told from the individualistic perspective, the history of the seventeenth century science appears rather distorted: the Scientific Revolution is often portrayed as though it originated with Copernicus in 1543 and concluded in 1687 with Newton. But this may only be true if we are interested in the opinions of individual scientists. I do not deny that this individualistic reading might be interesting for some purposes, but it definitely doesn't do justice to changes in the scientific mosaic and, therefore, is of little use as long as we are concerned with the assessment of a TSC. The history appears quite different when we focus our lenses on the level of the scientific mosaic. It is well known that at the level of the scientific mosaic the accepted theory of the world, for most part of the seventeenth century, was the good old Aristotelian natural philosophy (together with its medieval and early modern modifications and emendations, of course). The conclusion is that views of individual scientists can hardly be indicative of theory acceptance.[20]

But if not from individual confessions, how else can the historian determine what theories constituted the scientific mosaic at a given time? One can think of a number of possible indicators, such as *encyclopaedias, textbooks, university curricula*, and *minutes of association meetings*.[21] What all these sources have in common is that they normally represent the position not of this or that individual scientist, but of the whole scientific community. Encyclopaedias are typically good indicators of what was accepted at the time of their publication. As an outcome of collective effort, encyclopaedias usually provide us with a reasonably fair picture of what constituted the mosaic of the time. The same goes for academic textbooks. Although not always

---

[20] Shapin comes close to realizing this when he says that "even most educated people in the seventeenth century did not believe what expert scientific practitioners believed". Shapin (1996), p. 6.

[21] The idea that *textbooks* are indicators of acceptance can be found in Kuhn (1962/70), pp. 10, 136–138; Brush (1994), p. 136. That university *curricula* might be indicative of theory acceptance is acknowledged by Schmitt (1973), pp. 163, 176; Brockliss (1981), p. 35; Sturm and Gigerenzer (2006), p. 146.

as trustworthy as encyclopaedias, textbooks too are typically written with the objective of presenting the current state of knowledge in a particular field. When we open a contemporary physics textbook, for example, what we normally expect to find is an account of currently accepted physical theories, i.e. theories that the scientific community considers as the best available description of physical processes. In addition, we can refer to university curricula, since it is obvious that what we teach our students cannot be too different from what we take to be the case. Thus, by analyzing the university curricula of a particular time period we may find what theories were accepted at the time. Finally, minutes of association meetings can also be indicative of the stance of the community towards this or that theory.

None of these indicators is trouble-free, of course. One difficulty with taking encyclopaedias as evidence is that, until recently, encyclopaedias were sporadic phenomena, since the re-publication of even the most famous encyclopaedias could take decades. As a result, at least some of the content of printed encyclopaedias would very soon become obsolete. Therefore, encyclopaedias may be considered indicative of the scientific mosaic only at the time of their publication. Naturally, this leaves many "white spots" as far as the successive states of the scientific mosaic are concerned. Encyclopaedias alone cannot provide us with a thorough description of all the successive states of the mosaic.

Another problem with encyclopaedias is that, up until the eighteenth century, they were written by either an individual author (e.g. Pliny the Elder, Isidore of Seville) or an isolated group of scientists (e.g. the Brethren of Purity). This raises a question as to how representative these encyclopaedias were. Naturally, if an encyclopaedia is authored by an individual scientist it will most probably contain controversial claims championed by the author but not necessarily accepted by the community. Therefore, we should be cautious when dealing with encyclopaedias published before the mid-eighteenth century. Obviously, textbooks too are susceptible to similar drawbacks and must be treated with equal caution.

Taking university curricula as evidence has its own peculiar problems. A theory or an idea can often be taught at universities without being actually accepted as the best available description of its object. Take, for instance, the status of the Copernican theory at Wittenberg University in the second half of the sixteenth century. Some of the elements of the theory were presented merely as useful calculating tools, but not as the best available description of cosmos. Nowadays we have similar examples. We teach our students classical mechanics not because we take it to be the best available description of its domain, but because we still consider it extremely useful in practical applications. Therefore, when dealing with curricula, one must always clarify how exactly a particular theory was presented – was it introduced as accepted, as useful, or merely as pursuit-worthy?

In addition, we must bear in mind that there is a longstanding tradition of introducing students to views that are neither accepted, nor used, nor even pursued. Obviously, when nowadays we include Aristotelian or Cartesian natural philosophies in our courses on the history of science we do so not because we think these theories are acceptable, useful, or pursuit-worthy. We do so primarily out of historical interest. Therefore, when we come across a course on, say, Plato's natural philosophy taught in Italian universities in the sixteenth–seventeenth centuries, we shouldn't be confused: Plato's natural philosophy was presented as a matter of historical interest (and perhaps even as something worth pursuing), but certainly not as part of the mosaic of the time.[22]

Another common problem is that it usually takes some time before a newly accepted theory enters into encyclopaedias, textbooks, and university curricula. The average gap between the actual acceptance of a theory by the community and the respective correction of encyclopaedia articles, textbook chapters, and syllabi can differ from time period to time period. How long it normally takes for newly accepted parts of the mosaic to enter into encyclopaedias, textbooks, and curricula at different time periods is an important factual question which can only be tackled by HSC. Naturally, the historian must take this time gap into account and calibrate her conclusions accordingly. In addition, some of the most advanced of our accepted theories may never make it into university curricula or even into our encyclopaedias, simply because of their complexity, the narrowness of their scope, or some other particular reason.

In such cases, the minutes of respective association meetings can become indicative.[23] The problem with minutes is that in general they can provide only a very fragmentary picture of what was accepted and what was not. The fact that they normally focus on discussions of topical issues is both their virtue and their vice: they can help to clarify the stance of the community towards a very specific hypothesis, while saying little about the community's stance on many other theories. Thus, minutes can prove quite informative in some cases and virtually useless in others.

We should conclude, therefore, that none of these indicators is universal or conclusive. We have to concede that our reconstructions of the successive states of the mosaic will most probably remain incomplete. Establishing whether a theory was or wasn't accepted at time t may be quite simple in some cases and enormously difficult in others. While in some cases, such as the replacement of the Newtonian theory with that of Einstein, it may be possible to find out even the exact year of the replacement, in other cases it may turn out to be a virtually insoluble problem (e.g. when did al-Haytham's above-discussed reconciliation of Aristotelian cosmology with Ptolemaic astronomy become accepted?).

Nevertheless, we shouldn't become too pessimistic. In this respect, being confronted with the task of explicating the state of the scientific mosaic at a specific time, the historian of scientific change is in a position quite similar to that of the archaeologist or palaeontologist. They too suffer from chronic shortage of available data, and yet none of them takes this shortage of data as an insurmountable obstacle. Thus, we too shouldn't despair. At this point it is sufficient that we may know at least something. Yes, we lack detailed knowledge about the states of different mosaics before the late seventeenth century and we are yet to learn how to reconstruct these mosaics from incomplete data. In this sense, our "historical microscopes" have serious limitations. However, it is safe to say that the state of the mosaic during

---

[22] See Schmitt (1973), p. 163 and references therein.

[23] This indicator was suggested by Craig Knox during the seminar of 2013.

at least the last two or three centuries can in principle be obtained with reasonable precision. Taking the analogy further, we may even hope that, with the help of an accepted TSC, HSC will come up with increasingly sophisticated methods of reconstructing the successive states of the mosaic from seemingly unrelated data, just as archaeologists have learnt to reconstruct political, economic, religious, and social structures of early civilizations based on the examination of a limited number of survived artefacts.

As Michael Fatigati has shown in his pioneering work on Medieval Arabic scientific mosaic, indicators of acceptance can be quite different in different scientific communities and, importantly, there are ways of pinning them down. In particular, he has demonstrated that it is possible to take authoritative texts as reliable indicators of theory acceptance in Medieval Arabic scientific mosaic. As to which texts were considered authoritative, his suggestion is to track the so-called "license to teach [ijazah]" documents. This is because, in Medieval Arabic scientific mosaic, education was primarily done through explanation and memorization of authoritative texts and it is safe to say that the propositions in these authoritative texts were accepted as true. Thus, by pinning down theses authoritative texts, we can tackle the task of reconstructing Medieval Arabic scientific mosaic. Fatigati has also emphasized that two different mosaics may have two distinct sets of indicators of theory acceptance. Consequently, the historian may employ two different methods when trying to locate the elements of two different mosaics. It is a task of HSC to specify these methods for different scientific communities; there is good reason to believe that this goal is achievable.[24]

The next question that we shall address here concerns the *indicators of method employment*. How can the historian establish that a method was in fact employed at time *t*? Let us say that we are interested in finding out what method of appraisal was employed in natural philosophy in the early seventeenth century. Suppose, for instance, that according to some conjecture the method employed in theory appraisal in the early seventeenth century was a version of the hypothetico-deductivism. Question: how do we find out whether the employed method was in fact that of hypothetico-deductivism?

Traditionally, it has been assumed that in order to uncover the employed methods of a time period one should refer to the methodological proclamations of either individual scientists or the scientific community as a whole. To use the terminology clarified in section "Explicit and Implicit", it has been traditionally assumed that in order to unearth the method employed at the time one should refer either to the methodological proclamations of great individual scientists of that period or to the then-prescribed methodology. I shall consider these two strategies in turn.

One common strategy these days is to focus on the lines of reasoning of individual scientists. Take Anton Lawson's insistence that Galileo discovered Jupiter's

---

[24] The idea that indicators of theory acceptance and method employment are changeable has been suggested by Michael Fatigati and Joel Burkholder during the seminar of 2014. For his groundbreaking essay, Michael Fatigati has received the 2014 *Award for the Best Essay on the Theory of Scientific Change*. See Fatigati (2014).

moons by the hypothetico-deductive method and Douglas Allchin's criticism of Lawson's view. According to Lawson, Galileo's discovery of Jupiter's moons was hypothetico-deductive in nature, for by his observations Galileo was testing the so-called *moon hypothesis*. However, according to Allchin, this discovery was not hypothetico-deductive for Galileo didn't use hypothetico-deductive reasoning to predict Jupiter's moons.[25]

Now, as far as TSC is concerned, there are two major problems with this discussion. On the one hand, both authors conflate the individual and social levels. On the other hand, both authors fail to distinguish between method and methodology.[26] As I have explained in section "Explicit and Implicit", the question that is important from the standpoint of TSC is not what method the individual scientist employed, but what actually led to the replacement of one theory by another in the scientific mosaic. From this perspective, it makes no difference whatsoever what line of reasoning led Galileo to point his telescope at Jupiter. Similarly, from the perspective of TSC, it is not important whether Galileo's testimony does recount his original thinking. What is important is what respective changes to the scientific mosaic took place, when, and under what circumstances. What was the state of the scientific mosaic before the publication of *Sidereal Messenger*? Did the results of Galileo's observations become accepted by the community? More specifically, did proposition "Jupiter has moons" replace proposition "Jupiter has no moons" in the scientific mosaic? If so, when and why did the replacement take place? What effect did it have on other parts of the Aristotelian-medieval mosaic? It is questions like these that should be considered when we look at the process of scientific change from the perspective of TSC. Therefore, studying methodological views of individual scientists can hardly bring us any closer to explicating actual methods of science.

Another favourite strategy is to study methodologies openly prescribed by the scientific community. Some authors take the writings of the leading methodologists of the age and present a common denominator of their views as correctly describing the method employed at the time. This line of reasoning can be often found in the works of such grand masters as Lakatos and Laudan. In his account of the history of scientific method of the late 1700s and early 1800s, Laudan claims that there was a transition from the so-called *inductivist-empiricist* method to the so-called *method of hypothesis* (or hypothetico-deductive method, as we have it nowadays). In particular, it was a transition from the method which denied the legitimacy of hypothesizing about unobservable entities to the method which allowed the hypothesizing on the condition that hypotheses were subjected to empirical scrutiny. To justify that such a replacement "officially" took place, Laudan appeals to the works of Herschel and Whewell, great methodologists of the first half of the nineteenth century who recognized that the actual practice of science was at odds with the inductivist methodology openly prescribed at the time.[27]

---

[25] See Lawson (2002), Allchin (2003).

[26] There is also an obvious confusion of *construction* and *appraisal*, which can be ignored in this context.

[27] See Laudan (1984), pp. 55–59.

The flaw of this strategy is that it confuses two distinct phenomena – openly prescribed methodologies and actually employed methods. It is one thing to say that a methodology was prescribed by the scientific community at time t, and it is a completely different thing to claim that a method was actually employed in theory assessment at time t. Laudan himself seems to understand that the requirements of the openly prescribed methodology may or may not coincide with the requirements of the actually employed method. The case of the "official methodology" of the eighteenth century is an example which Laudan himself discusses in detail.[28] It is therefore not in the official proclamations of the community that we shall locate the actually employed methods. Of course, it never hurts to know the requirements of the methodology prescribed at the time, for sometimes these requirements can give us a useful hint on the actual expectations of the community. Yet, the point is that these openly prescribed requirements should not substitute for the actual expectations of the community of the time. For this reason, the list of primary indicators of method employment doesn't include textbooks or encyclopaedias. From textbooks and encyclopaedias it is possible to learn only what was openly formulated at the time and, thus, textbooks and encyclopaedias can help us to locate the methodology openly prescribed at the time. But in order to locate the employed method, we need other strategies.[29]

One correct strategy is suggested by the very definition of *employed method*: a method is said to be employed if theories become accepted only when their acceptance is permitted by the method. Therefore, what we have to do is track the transitions in the scientific mosaic of the time and try to reveal why these transitions took place. For example, if we want to unearth the method employed in natural philosophy in the early seventeenth century, we must consider the changes in the mosaic which took place in that period. We shall try to locate a theory (or even a single proposition) that became accepted sometime in the early seventeenth century and try to understand what implicit requirements it satisfied. If it turns out, for instance, that all of the transitions of that period took place when some of the novel predictions of proposed theories became confirmed, then this will be a good indicator that the actual method employed at the time was indeed along the lines of hypothetico-deductivism. But if it turned out that the transitions of the period took place without any novel predictions whatsoever, this would suggest that the actual method of the time wasn't quite in accord with the principles of hypothetico-deductivism. In other words, only a careful study of successive states of the mosaic can reveal what methods were employed at different periods.

Consider an example. It is well known that when the first telescopes were constructed in the early seventeenth century, their reliability was debatable. It was argued that although the telescope was a reliable tool for observing terrestrial

---

[28] See Laudan (1984), p. 54.

[29] In this, I disagree with Brush who apparently holds that open proclamations are indicative of the actual method. See his discussion of the acceptance of Mendeleev's theory in Brush (1994), p. 140. That textbooks often provide misleading information regarding employed methods was pointed out in Kuhn (1962/70), p. 137.

objects, it was not reliable for celestial observations. This dichotomy had to do with the then-accepted Aristotelian-medieval belief that terrestrial and celestial regions are completely distinct and what is true for one should not necessarily hold for the other. However, sometime after 1611, the legitimacy of the telescope as a tool for celestial observations was accepted and astronomical data obtained by means of the telescope began to be considered trustworthy.[30] Obviously, here we have an instance of change in the employed method, for the community clearly changed its expectations concerning astronomical observations. But when exactly did this transition take place? To answer this question, it isn't sufficient to establish when, how, and who first came up with the idea of studying the heavens with the telescope. Determining the year when the first results obtained with the telescope were published is equally irrelevant. What needs to be established is when exactly the results of telescopic observations began to be considered trustworthy. In particular, we have to ask: when did a telescopic observation lead to changes in the mosaic for the first time, that is, to the acceptance or rejection of a theory, or even of a single proposition, such as "Jupiter has moons"?[31] Was it immediately after the publication of *Sidereal Messenger*? Was it during Galileo's lifetime or sometime after his death? Or was it only when geocentrism implicit in the Aristotelian natural philosophy was replaced by the conception of heliocentric solar system and infinite universe that was part of the Cartesian natural philosophy? Answering questions of this sort is vital for uncovering what methods were employed at a certain period and how they changed throughout time.

Finally, a few words must be said about available contenders. How do we know the list of proposed modifications to the mosaic at time *t*? Put differently, how can we clarify the list of contender theories available at the time? Unlike the question of states of the mosaic, the answer to this question is quite straightforward. After all, if there is one aspect of the history of science that we seem to know pretty well it is when, how, and by whom scientific theories were constructed. There is no shortage of historical studies focused on the circumstances of theory construction. In fact, we can name virtually any year (at least from the modern period onward) and the historian will swiftly come up with a comprehensive list of theories and ideas available at the time. In other words, the list of the modifications proposed at a specific time is readily obtainable.

This concludes *Part I* where I have attempted to define the *scope* of TSC, to show that it is *possible* and to clarify how it is to be *assessed*. Now that we know what TSC is and how it is possible, I shall move to the second part of the project – the construction of an actual TSC.

---

[30] See Van Helden (1994).

[31] Note that this is not the same as asking when it was the first time that a telescopic observation has made us consider one theory more *useful* than another. See section "Acceptance, Use, and Pursuit" for details.

Part II
Theory

Chapter 4
Axioms

The two classes of elements that can undergo scientific change are accepted *theories* and employed *methods*. Therefore, the laws of scientific change must cover changes in both theories and methods. I shall first posit the four laws that I believe govern the process of scientific change. I realize that, at this stage, some of the laws may seem unclear or disputable. The only remedy, I am afraid, is to take a piecemeal approach and clarify the laws by means of relevant historical illustrations as well as the theorems that follow from them. Thus, I shall consider each of these four laws in turn.

The First Law: Scientific Inertia

According to *the first law*, any element of the mosaic of accepted theories and employed methods remains in the mosaic except insofar as it is overthrown by another element or elements. Basically, the law assumes that there is certain inertia in the scientific mosaic: once in the mosaic, elements remain in the mosaic until they get replaced by other elements. It is reasonable therefore to call it *the law of scientific inertia*.

The first law has two obvious corollaries: as there are two classes of elements in the mosaic, the first law can be specified for theories and for methods. Let us start with *the first law for theories*:

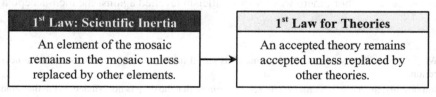

1ˢᵗ Law: Scientific Inertia	1ˢᵗ Law for Theories
An element of the mosaic remains in the mosaic unless replaced by other elements.	An accepted theory remains accepted unless replaced by other theories.

© Springer International Publishing Switzerland 2015
H. Barseghyan, *The Laws of Scientific Change*, DOI 10.1007/978-3-319-17596-6_4

When applied to theories, *the first law* stipulates that an accepted theory is not rejected unless there is a suitable replacement.[1] If for some reason scientists of a particular field stop pursuing new theories (i.e. stop producing new alternatives) the last accepted theory will safely continue to maintain its position in the mosaic. Similarly, no "further confirmations" are required in order to keep an accepted theory accepted. Once a theory becomes accepted, there is no need for additional arguments in its favour.

Importantly, *the first law for theories* does not specify what kind of theories can replace an accepted theory. At minimum, an accepted proposition can be replaced by its own negation. Suppose, the scientific community in question accepts that a certain drug is therapeutically efficient in alleviating a certain condition. In principle, this proposition can be replaced by its own negation, i.e. the proposition that the drug is not efficient in alleviating the condition.[2] HSC shows many examples of this sort. Recall, for instance, the medieval and early modern belief that bloodletting is efficient in restoring the proper balance of humors in the body and, thus, restoring health. When this belief was rejected it was simply replaced by its negation.

What is important, however, is that this replacement-by-negation scenario is by no means universal; we don't find it in every field of science and at every time period. It is a historical fact that many scientific communities have imposed additional requirements on what sort of propositions can replace the accepted propositions. In the contemporary physics, for example, a mere negation of an accepted theory would not be sufficient. Consider, for instance, the case of quantum field theory (QFT) in the 1950–60s. In the late 1940s, QFT was successfully applied to electromagnetic interactions by Schwinger, Tomonaga, and Feynman when a new theory of quantum electrodynamics (QED) was created. Hope was high that QFT could also be applied to other fundamental interactions. However, it soon became apparent that the task of creating quantum field theories of weak and strong interactions was not an easy one. It was at that time (the 1950–60s) when QFT was severely criticized by many physicists. Some physicists criticized the techniques of renormalization which were used to eliminate the infinities in calculated quantities. Dirac, for instance, thought that the procedure of renormalization was an "ugly trick". Another line of criticism was levelled against QFT by Landau, who argued in 1959 that QFT had to be rejected since it employed unobservable concepts such as local field operators, causality, and continuous space time on the microphysical level. It is a historical fact however that, all the criticism notwithstanding, QFT was not rejected.[3] In short, there was serious criticism levelled against the then-accepted

---

[1] What makes a theory "suitable replacement" is stated in *the second law*, discussed in the next section.

[2] Importantly, this is not a return to state zero: to believe that the drug is not efficient is not the same as to have no belief on the subject whatsoever. To say that we don't know whether the drug is efficient is one thing; to say that it is inefficient is quite another. The former has no propositional content, while the latter clearly contains some knowledge, albeit negative.

[3] See Kragh (1999), pp. 332–348.

theory, but it didn't lead to its rejection, for the physics community of the time didn't allow for a simple replacement-by-negation scenario.[4]

Moreover, even a considerable body of empirical evidence against an accepted physical theory might not be sufficient. In fact, we would be prepared to reject our accepted physical theory only if there were another physical theory on the market that, among other things, explained, by and large, the facts explained by the currently accepted theory. Thus, when it comes to empirical theories, nowadays we do not reject our accepted empirical theories even when these theories face anomalies (counterexamples, disconfirming instances, unexplained results of observations and experiments). This anomaly-tolerance has been a feature of empirical science for a long time.

The famous case of Newtonian theory and Mercury's anomalous perihelion is a good indication that anomalies were not lethal for theories also in the nineteenth century empirical science. In 1859, it was observed that the behaviour of planet Mercury doesn't quite fit the predictions of the then-accepted Newtonian theory of gravity. The rate of the advancement of Mercury's perihelion (precession) wasn't the one predicted by the Newtonian theory. For the Newtonian theory this was an anomaly. Several generations of scientists tried to find a solution to this problem. But, importantly, this anomaly didn't falsify the Newtonian theory. The theory remained accepted for another 60 years until it was replaced by general relativity circa 1920.

This wasn't the first time that the Newtonian theory faced anomalies. In 1750 it was believed that the Earth is an oblate-spheroid (i.e. that it is flattened at the poles). This was a prediction that followed from the then-accepted Newtonian theory, a prediction that had been confirmed by Maupertuis and his colleagues by 1740. However, soon very puzzling results came from the Cape of Good Hope: the measurements of Nicolas Louis de Lacaille were suggesting that, unlike the northern hemisphere, the southern hemisphere is prolate rather than oblate.[5] Thus, the Earth was turning out to be pear-shaped! Obviously, the length of the degree of the meridian measured by Lacaille was an anomaly for the accepted oblate-spheroid view and, correspondingly, for the Newtonian theory. Of course, as with any anomaly, this one too forced the community to look for its explanation by rechecking the data, by remeasuring the arc, and by providing additional assumptions. Although it took another 80 years until the puzzle was solved, Lacaille's anomalous results didn't lead to the rejection of the then-accepted oblate-spheroid view. Finally, in 1834–38,

---

[4] Criticism, to be sure, may lead to the *construction* of new theories (*the first law* doesn't impose limitations in this regard). In this sense, criticism can be quite fruitful. After all it is the dissatisfaction with currently accepted theories that often forces scientists to look for alternative explanations. But as far as the *acceptance/rejection* of theories is concerned, pointing out the drawbacks of accepted theories is not enough. An accepted proposition remains in the mosaic unless it is replaced by some other proposition. It remains to be seen how exactly an accepted theory can be replaced; what is clear is that it is never replaced with nothing. The mechanism of theory rejection is discussed in section "Rejection of Elements".

[5] For a discussion of Lacaille's measurements, see Evans (1967) and Evans (1992). An account of Lacaille's observations is also provided in Maclear (1866).

Thomas Maclear repeated Lacaille's measurements and established that the deviation of Lacaille's results from the oblate-spheroid view were due to the gravitational attraction of Table Mountain.[6] The treatment of Lacaille's results – as something bothersome but not lethal – reveals the anomaly-tolerance of empirical science even in the eighteenth century.[7]

HSC provides us with even earlier examples of this phenomenon. Take the Aristotelian-medieval natural philosophy accepted up until the late seventeenth century. Tycho's Nova of 1572 and Kepler's Nova of 1604 seemed to be suggesting that, contrary to the view implicit in the Aristotelian-medieval mosaic, there is, after all, generation and corruption in the celestial region. In addition, after Galileo's observations of the lunar mountains in 1609, it appeared that celestial bodies are not perfectly spherical in contrast to the view of the Aristotelian-medieval natural philosophy. Moreover, observations of Jupiter's moons (1609) and the phases of Venus (1611) appeared to be indicating that planets are much more similar to the Earth than to the Sun in that they too have the capacity for reflecting the sunlight. All these observational results were nothing but anomalies for the accepted theory which led to many attempts to reconcile new observational data with the accepted Aristotelian-medieval natural philosophy. What is important is that the theory was not rejected; it remained accepted throughout Europe for another 90 years and was overthrown only by the end of the seventeenth century.

The abundance of similar historical cases is arguably the main reason why Kuhn, Lakatos, Laudan and many others have rejected Popper's *falsificationist* view that the whole course of science is nothing but a series of conjectures and their refutations.[8] Although there have been many followers of falsificationism among scientists,[9] nowadays it is commonly accepted among philosophers of science that counterexamples do not kill theories. Following Duhem's pioneering work, Kuhn was among the first to highlight that anomalies are not something exceptional and

---

[6]A detailed account of Maclear's operation is his (1866). For discussion, see Evans (1958).

[7]One sound indication of this fact is found in d'Alembert's article *Figure de la Terre* in *Encyclopédie*, in which d'Alembert presents Lacaille's results as an anomaly – but, interestingly, not even as an anomaly for the oblate-earth view. The only thing that seems to be concerning d'Alembert is that the results deviated *quantitatively* from the value of the Earth's ellipticity calculated on the basis of the results of earlier expeditions. The view that the Earth is an oblate spheroid is never even questioned. See Diderot (ed.) (1751–1780), vol. 6, pp. 755–756.

[8]As Lakatos explained, Popper's actual position should be differentiated from that of dogmatic falsificationism, for Popper did realize that in reality general theories cannot be *decisively* refuted by counterexamples. He knew that singular propositions describing observations and experiments are as fallible as general theoretical propositions and, thus, the fault may lay not only with the theory but also with the observation. See Lakatos (1970), pp. 12–47 for his typology of falsificationisms – *dogmatic, naïve methodological*, and *sophisticated*. In the appendix, Lakatos explains why Popper is not a dogmatic falsificationist. See pp. 93–96.

[9]See, for instance, Medawar (1979), pp. 86–88. It is unfortunate that some popular science writers still seem to openly accept falsificationism. Take, for example, the following passage by physicist Michio Kaku: "Needless to say, Einstein's theory has withstood the test of time for almost a century and if there's one data-point out of place, we would have to throw the entire theory out." Kaku (2011). Needless to say, we would *not*.

that normally the mere presence of "refuting" counterexamples does not lead to theory rejection.[10] Lakatos too recognized that, for any theory, there is always "an ocean of anomalies", but these anomalies play a role in theory acceptance/rejection only when they happen to confirm some novel predictions of an alternative theory. In the absence of an alternative theory, they never lead to the rejection of an accepted theory. As Lakatos puts it, there are no negative crucial experiments.[11] Similarly, according to Laudan, "it is not true that, in general, the discovery of an anomaly for a particular theory will lead, in and of itself, to the abandonment of the theory which exhibits the anomaly."[12]

Yet, it needs to be appreciated that this anomaly-tolerance is by no means a universal feature of science. There are both historical and theoretical reasons to believe that the attitude of the community towards anomalies is historically changeable and non-uniform across different fields of science.

A quick glance at the historical record reveals that there have been both anomaly-tolerant and anomaly-intolerant attitudes. While it is safe to say that the modern empirical science is anomaly-tolerant (i.e. counterexamples do not lead to theory rejection in contemporary physics, chemistry, or biology), it is also clear that the fields of formal science (e.g. logic, mathematics) are anomaly-intolerant. Consider the famous *four color theorem* currently accepted in mathematics which states that no more than four colors are required to color the regions of the map so that no two adjacent regions have the same color. Suppose for the sake of argument that a map were found such that required no less than five colors to color. Question: how would mathematicians react to this anomaly? Yes, they would check, double-check, and triple-check the anomaly, but once it were established that the anomaly is genuine and it is not a hoax, the proof of the four color theorem would be revoked and the theorem itself would be rejected. Importantly it could be rejected without being replaced by any other *general* proposition. Its only replacement in the mosaic would be the *singular* proposition stating the anomaly itself. This anomaly-intolerance is a feature of our contemporary formal science.[13] Thus, we have to accept that anomaly-tolerance is not a universal feature of science.

This conclusion is also supported theoretically. Indeed, *the first law for theories* doesn't impose any limitations as to what sort of propositions can in principle replace the accepted propositions; it merely says that there is always *some* replacement. This replacement can be as simple as a straightforward negation of the accepted proposition, or a full-fledged general theory, or a singular proposition

---

[10] See Kuhn (1962/70), p. 81, 84–87.

[11] See Lakatos (1971), pp. 111, 126–128.

[12] Laudan (1977), p. 37.

[13] A case can be made that the contemporary analytic philosophy is another example of anomaly-intolerance. Analytic philosophers often take a non-compromise approach when explicating the meaning of an obscure concept; one disconfirming intuition is commonly taken as reason enough to void the results of years of meticulous analysis. However, one can also claim that this historical hypothesis doesn't hold water, since in order to be rejected something needs to be accepted first and there is virtually nothing accepted in analytic philosophy. This is an interesting historical issue and must be tackled by HSC.

describing some anomaly. The actual attitude of the community may be different at different time periods and in different fields of science. As to what determines the actual expectations of the community and how they change, the answer will become obvious once we study *the third* and *the zeroth laws* of scientific change.

Meanwhile, let us quickly consider the second corollary of *the law of scientific inertia – the first law for methods*:

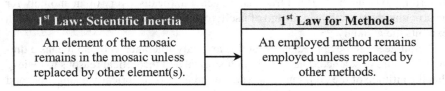

1st Law: Scientific Inertia	1st Law for Methods
An element of the mosaic remains in the mosaic unless replaced by other element(s).	An employed method remains employed unless replaced by other methods.

Formulated for methods, the first law simply says that the implicit expectations employed in theory assessment will continue to be employed unless they are replaced by some alternative expectations. Several authors have expressed this idea in one way or another. In his *Reflections on my Critics*, Kuhn wrote: "scientists behave in the following ways; those modes of behaviour have... the following essential functions; in the absence of an alternative mode *that would serve similar functions*, scientists should behave essentially as they do if their concern is to improve scientific knowledge."[14] If we disregard the normative element of Kuhn's formulation ("scientists should behave"), this is essentially Kuhn's expression of *the first law for methods*: insofar as there are no alternative methods we stick to our current requirements.[15]

Again, the law doesn't impose any limitations on the sort of methods that can replace the employed methods. In the most basic case, the community can reject some of the more specific requirements of the currently employed method and revert to a more abstract method. Or it can replace those rejected requirements with some new specific requirements. Suppose the employed method stipulates that a new theory must be tested in repeatable experiments and observations. In principle, the community may one day remove some of the ingredients of this method, say, the requirement of repeatability. As a result, the community can either revert to a more abstract method or it can introduce a new requirement to replace the repeatability clause. For instance, the community may revert to the more abstract method which stipulates a new theory must be tested in experiments and observations (no repeatability requirement). Alternatively, it can introduce a new requirement that in addition to empirical testing a new theory must also explain all the facts explained by the accepted theory. Which of these two scenarios materialize at each particular instance is decided by a number of contingent factors.[16] In any case, what the first law says

---

[14] Kuhn (1970a), p. 237 (Kuhn's *emphasis*).

[15] See section "Descriptive and Normative" for discussion. This confusion of normative and descriptive, typical for Kuhn, has led Nola and Sankey to interpret this passage as pertaining to metamethodology. See Nola and Sankey (2000), p. 29.

[16] See section "Rejection of Elements" for details.

is that the community never remains with no expectations whatsoever. When facing a new theory, the community always has *some* implicit expectations concerning such theories. These expectations may be very specific or they may be very abstract and vague, but *some* expectations are always present, for otherwise no theory assessment would be possible. That is essentially the point of *the first law for methods*.

To conclude this section, let us sum up *the first law* and its corollaries. *The law of scientific inertia* essentially states that an element of the mosaic is never given up for nothing, i.e. accepted theories and employed methods remain in the mosaic unless replaced by other theories or methods. The law doesn't impose any limitations as to what sorts of theories and methods can in fact replace the accepted theories and employed methods; these might be different at different time periods and in different fields of science.

The Second Law: Theory Acceptance

While *the first law for theories* only says how theories *remain* in the mosaic, *the second law* tells us how theories *become* accepted. *The law of theory acceptance*, as we can also call it, states that theories become accepted only when they satisfy the requirements of the methods actually employed at the time. In other words there is only one way for a theory to become accepted – it must meet the implicit expectations of the scientific community. As I indicated earlier, this law is a direct consequence of a definition of *employed method*.[17] Since *employed method* is defined as a set of implicit criteria actually employed in theory assessment, it is obvious that any theory that aims to become accepted must meet these requirements.

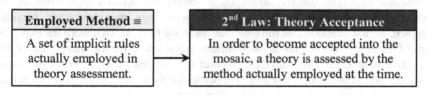

Employed Method ≡	**2nd Law: Theory Acceptance**
A set of implicit rules actually employed in theory assessment.	In order to become accepted into the mosaic, a theory is assessed by the method actually employed at the time.

Therefore, *the second law* merely explicates what was already implicit in our definition of *employed method*.[18] Of course, a theory may sometimes appear as though it became accepted in violation of the method employed at the time, but it may only *appear* so. In reality, a theory may violate the *methodology* to which the

---

[17] See section "Explicit and Implicit".

[18] So, properly speaking, the second law is not a law in the traditional sense, for normally a law is supposed to have some empirical content, i.e. its opposite should be conceivable at least in principle. Obviously, the second law is a *tautology*, since it follows from the definition of *employed method*. My only excuse for presenting it as a law is that it is too important to be lost in the thickets of definitions.

scientific community openly subscribes, but not the actually employed method. The actual expectations (i.e. the method) cannot be violated, for if they were, they wouldn't be the *actual* expectations! If it appears that a theory became accepted in violation of the requirements of the time, this will be a clear-cut indication that our *knowledge* of the actual requirements employed at the time is deficient. Methods cannot be violated (by definition), only methodologies can. As I explained in section "Indicators", we can only learn about our actual expectations by analyzing transitions in our mosaics, i.e. transitions from one accepted theory to the next. So, when a theory becomes accepted, the question that we should ask is not whether the theory violated our actual requirements (for it obviously couldn't), but rather what the actual requirements that allowed the theory to become accepted were.

Even the most "revolutionary" theories must meet the actual requirements of the time in order to become accepted. Einstein's general relativity is considered as one of the most ground-breaking theories of all time and, yet, it was evaluated in an orderly fashion and became accepted only after it satisfied the requirements of the time. From that episode we can reconstruct what the actual requirements of the time were. It is well known that the theory became accepted circa 1920, after the publication of the results of Eddington's famous observations of the Solar eclipse of May 29, 1919 which confirmed one of the novel predictions of general relativity – namely, the deflection of light in the spacetime curved due to the Sun's mass. Thus, it is safe to say that the scientific community of the time expected (among other things) that a new theory must have confirmed novel predictions.[19]

In a similar fashion, we can try to reconstruct actual expectations of scientific communities of different time periods and different fields. Suppose we study the history of the transition from the Aristotelian-medieval natural philosophy to that of Descartes in France and that of Newton in Britain circa 1700. It follows from *the second law* that both theories managed to satisfy the actual expectations of the respective scientific communities, for otherwise they wouldn't have become accepted. So the question that the historian must ask here is: what were the expectations of the respective scientific communities that allowed for the acceptance of the respective natural philosophies? *The second law* suggests that, in order to reconstruct the actual method employed at a particular time, we must study the actual transitions in theories that took place at that time.

Obviously, these expectations may be different for different types of theories. By the second law, in order to become accepted, a physical theory should conform to the implicit requirements of the community regarding physical theories, i.e. to the method of physics. In order to become accepted, a sociological theory must meet the implicit expectations of the community regarding sociological theories, i.e. the method of sociology. The same goes for any other theory in any other field of science: in order to become accepted, a theory must satisfy the community's implicit expectations concerning theories of that type (whatever those expectations might be at the time of assessment).

---

[19] Ernan McMullin sounds along these lines in McMullin (1988), p. 35.

HSC faces an extremely interesting (albeit enormously challenging) task of reconstructing those implicit expectations. It wouldn't be too much of an exaggeration to say that this issue has yet to be tackled by HSC, since to date *implicit* requirements regarding theories in different fields of inquiry and different time periods have been chiefly overlooked. And this is not a criticism of the historian; this is simply an indication of how many interesting issues come to light when historical research is guided by a proper theory. For one, this issue could not have arisen, had we not clearly distinguished between *method* (implicit expectations) and *methodology* (openly prescribed requirements). In addition, the question could not have arisen, had there been no *second law* that clearly states the mechanism of theory acceptance and the role of the community's implicit expectations in the process. With the *method/methodology* distinction and *the second law* at hand, the historian can delve into the respective historical contexts and try to bring to light the actual expectations of the community regarding different types of theories.

The second law has another noteworthy consequence. Consider the famous debate on the status of novel predictions. This debate has deep roots and can be traced back at least to Whewell and Mill. The question at issue, in its most general form, is this: is it sufficient for a new theory to explain only known facts in order to be accepted or should it necessarily have confirmed novel predictions? While authors like Popper, Lakatos, or Musgrave argue for a special status of novel predictions, others like Hempel, Carnap, or Laudan maintain that as far as theory assessment is concerned there is no substantial difference between novel predictions and post factum explanations of known facts (sometimes called *retro*-dictions). Those who believe that novel predictions are evidentially advantageous proceed to the second question about the nature of *novelty*: should it be understood as temporal novelty, as heuristic novelty, or as something else?[20]

It can be shown that the whole debate in its current shape is ill-founded.[21] In particular, it follows from *the second law* that we cannot simply ask whether a theory needs confirmed novel predictions in order to become accepted. We must specify the question: what *time periods* and what *fields of inquiry* are we interested in? Whether a new theory is expected to have confirmed novel predictions in order to become accepted is, according to *the second law*, decided by the *method* employed at the time. It is conceivable that at some time periods and in some fields of inquiry new theories were supposed to provide confirmed novel predictions while at other time periods and in other fields of inquiry they were not. In short, it would be too naïve to assume that confirmed novel predictions were expected from *all* theories in *all* fields and at *all* time periods. Thus, we may rightfully ask whether physical theories were

---

[20] See footnote 5 on page 101 for references.

[21] It is also ill-founded in a more straightforward way: the question that both parties seem to be discussing is a mixture of descriptive and normative issues. In fact, there is not one but two questions merged together. It is one thing to ask whether a theory *is* actually required to have confirmed novel predictions in order to become accepted (*descriptive* question) and it is another thing to ask whether new theories *ought to* be required to have confirmed novel predictions (*normative* question). Obviously, only the descriptive question concerns us here. See section "Descriptive and Normative".

required to have confirmed novel predictions in the 1880s, or whether astronomical theories in the 1670s were supposed to have confirmed novel predictions, or whether sociological theories nowadays are expected to have any confirmed novel predictions. Asking whether novel predictions are a must without specifying what time period and what fields of inquiry we are interested in simply does not make any sense. This is an immediate consequence of *the second law*.

In brief, the second law suggests that theories become accepted only if their acceptance is permitted by the method employed at the time of the assessment. This law follows from the very definition of employed method. This holds for any theory in any field of science: to become accepted any theory must meet our implicit expectations regarding theories of that type. It is a task of HSC to reconstruct those implicit expectations for different fields of science at different time periods.

The Third Law: Method Employment

Unlike the second law, *the third law* is far from trivial. In a sense, it is the most central of all four laws, for it governs transitions from one employed method to the next. Thus, it can be called *the law of method employment*[22]:

3rd Law: Method Employment
A method becomes employed only when it is deducible from other employed methods and accepted theories of the time.

Before we delve into technicalities, it pays to put the third law in more simple language. Essentially, *the third law* stipulates that our accepted theories shape our employed methods. The basic idea that our knowledge of the world greatly affects our employed methods has been appreciated since the 1970s and 1980s by Kuhn, Feyerabend, Shapere, Laudan, and McMullin.[23]

This idea plays an important role in the later Laudan's *Science and Values*. Laudan clearly recognizes that "the proper procedures for investigating the world have been significantly affected by our shifting beliefs about how the world works."[24] His favourite example illustrating this point is that of the impact of our knowledge

---

[22] At first sight, the third law may give an impression that it's too strong and gives no room for real innovation. Yet, that impression is false. For discussion, see sections "Scientific Underdeterminism" and "The Role of Methodology" below.

[23] See Kuhn (1962/70), p. 109.

[24] Laudan (1984), p. 39. However, we must bear in mind that Laudan doesn't quite distinguish between *method* and *methodology*. As a result, he often seems to be saying that accepted theories shape our *methodologies*, which is not the same as saying that our accepted theories shape our *methods*. See section "Explicit and Implicit".

about the placebo effect and the experimenter's bias on our methods of drug testing. Our procedures for testing the efficacy of a drug have gradually evolved as we have learnt that the improvement in a medical condition can be due to unaccounted effects (e.g. improved nutrition), the placebo effect or the experimenter's bias. As a result, nowadays the practice of drug testing requires an implementation of the so-called *double-blind trials*: neither patients nor researchers who are in contact with the patients must know which of the two groups is the active group and which is the control group. In short, our drug testing methods have changed as a result of changes in our knowledge of the world.

Another example of a method being shaped by our knowledge of the world has been pointed out by McMullin. In the early eighteenth century, there was a transition from the Aristotelian requirements to "the method of hypothesis" (the hypothetico-deductive method, as we call it nowadays). As McMullin has explained, the transition had to do with the recognition of the fact that the world is much more complex than it appears in observations, i.e. that it contains underlying causal mechanisms that produce all the observable effects. Once we have appreciated this fundamental idea, it changed our method of theory appraisal: from that point on, an explanatory hypothesis about inner workings of the world was supposed to be scrutinized by testing the novel predictions that followed from it.[25]

Examples of methods being shaped by theories are abundant not only in the modern, but also in the ancient and medieval science. David Lindberg provides several illustrations of this phenomenon. Take Plato's conviction that the true knowledge is to be gained through reason alone. It is readily seen that this require-ment is based on a number of assumptions concerning man and nature. In particular, it is based on Plato's belief that the senses are incapable of revealing the true reality and that the truth is nevertheless achievable. The former premise, in turn, is based on deeper convictions that the senses reveal only physical objects which are merely imperfect replicas of eternal ideas, and that these eternal ideas alone constitute the true reality.[26]

Thus, the general idea of *the third law* is not new.[27] Yet, it would be fair to say that the process itself has been understood only in outline. The actual details of *how* accepted theories shape employed methods have not been explained with necessary precision. What we have had so far is a picture from a bird's eye perspective. What we lack is the knowledge of the actual mechanism: how exactly can accepted theories shape employed methods? In particular, do employed methods *logically follow* from accepted theories, or do they merely *cohere* with accepted theories, or is there some other relation between the two? *The third law* is my attempt to fill this gap by explaining how exactly accepted theories affect employed methods. I shall first explain the mechanism of method employment by scrutinizing a well-studied case – Laudan's favourite case of drug testing methods. After that I will show that the same mechanism also applies to any other instance of method employment, be

---

[25] See McMullin (1988), pp. 32–34. Feyerabend had hinted this earlier in his (1975), pp. 232–233.

[26] See Lindberg (2008), pp. 35–37; see also p. 34 for the cases of Parmenides and Democritus.

[27] Among contemporary authors who appreciate this idea is Brown (2001), pp. 137–140.

that a very specific method (e.g. cell counting methods) or a very general method (e.g. the Aristotelian-medieval method, the hypothetico-deductive method, etc.).

Say there is a new drug for alleviating depression. Question: how do we test it? The simplest way of testing the therapeutic efficacy of the drug is to administer it to a group of patients suffering from depression and to check the outcome. If there is a noticeable improvement in the condition of the patients, then we conclude that the drug is efficient. However, as Avicenna has pointed out, an improvement in a medical condition can be due to many unaccounted effects, such as a body's natural healing ability, improved nutrition and so on. So how can we ensure that the improvement was due to the drug itself and not due to other unaccounted factors? The *controlled trial* was Avicenna's answer. We organize a trial with two groups of patients with the same condition – the active group and the control group. Only the patients in the active group receive the drug. The drug is said to be therapeutically efficient only if the improvement in the active group is greater than in the control group. In principle, this should help to minimize the effect of unaccounted factors.

What we have here is a transition from one method to another triggered by a new piece of knowledge about the world. The initial method was something along the lines of hypothetico-deductivism: we had a hypothesis "the drug is effective in alleviating depression" and we wanted to confirm it experimentally. Once we learnt that the alleviation may be due to other factors, our initial method was modified to require that a drug's efficacy must be tested in a controlled trial:

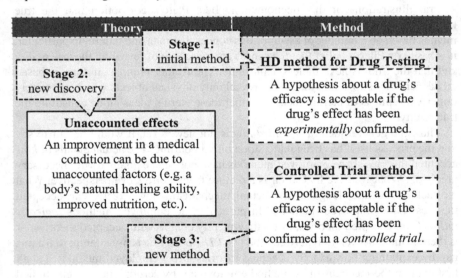

A few words about the above diagram are necessary here. Since we deal with *methods* (and not *methodologies*), we must bear in mind that any attempts to explicate them are provisional. When I try to explicate some method presumably employed at a certain time period, I realize that my explication itself is a fallible historical hypothesis. I suggest we present our explications of methods in dashed rectangles to indicate that in reality we deal with implicit requirements and that our explications of these methods may or may not be correct. Also, observe that at the moment there are no arrows in the diagram. The absence of arrows indicates that we do not yet know the exact mechanism of how theories "shape" methods, so we are not yet in a position to draw any arrows connecting the initial method, the new theory, and the new method. I shall add the missing arrows, once we unearth the mechanism of the process.

In order to do that, I shall consider another transition in the method of drug testing which took place when we learned about the so-called *placebo effect*, i.e. that the improvement in patients' condition can be due to the patients' belief that the treatment will improve their condition. This new knowledge obviously forced us to alter the drug testing procedure. It was no longer sufficient to have two groups of patients. If only one of the two groups received the drug then the resulting positive effect could be due to the patients' belief that the drug was really efficient in alleviating their condition. The solution was to organize a *blind trial*. We take two groups of patients with the condition, but this time we make sure that both groups of patients believe that they undergo treatment. However, only the patients of the active group receive the real drug; to the patients in the control group we give a placebo (fake treatment), but tell them that they undergo a real treatment. If the improvement in the active group is greater than in the placebo group, then we conclude that the drug is efficient. Again, this is an instance of a method change brought about by a change in accepted theories:

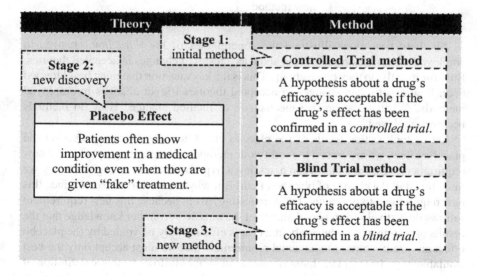

Finally, the drug testing method changed again when we discovered the phenomenon of *experimenter's bias*, i.e. the fact that researchers involved in drug testing can influence the outcome of the tests. In particular, as we have learnt during the last 50 years, the researchers that are in contact with patients can give patients conscious or unconscious hints as to which group is which. It is possible that the positive effect of the drug established in a blind trial was due to the fact that the patients in the placebo group knew that they were given a placebo. The method of drug testing was modified yet again to reflect this newly discovered phenomenon. The contemporary approach is to perform a *double-blind trial* where neither patients nor researchers know which group is which. This is another instance of method change:

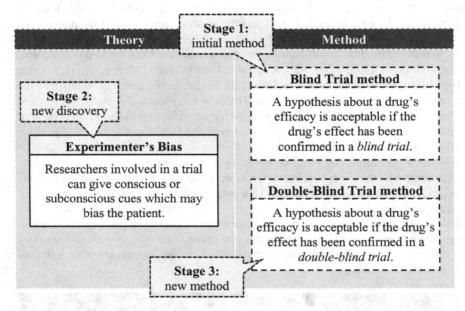

All these transitions illustrate the general idea of *the third law*: changes in employed methods normally come about as a result of changes in accepted theories. But how exactly do they come about? This is the key question that must be addressed here: how exactly can changes in accepted theories trigger changes in employed methods? What is the precise mechanism of method change? How do methods become employed?

Let us look at the details of the placebo effect episode. Once we discover the placebo effect, we are forced to modify our method of drug testing by adding a new requirement – namely, that when assessing a hypothesis about a drug's efficacy we must forestall the chance of placebo effect. But why are we forced to introduce this new requirement to our method of drug testing? Well, because this new requirement follows deductively from two elements of the mosaic – from our knowledge that the results of testing a hypothesis about a drug's efficacy may be voided by the placebo effect and from a more fundamental requirement that we must accept only the best available hypotheses. (This latter requirement simply follows from the definition of *acceptance*: to accept something is to take it as the best available description of its object.[28]) Here is the detailed deduction:

---

[28] See section "Acceptance, Use, and Pursuit" for details.

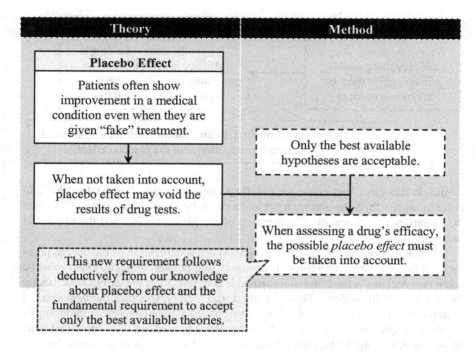

Note, however, that this new requirement is not identical with *the blind trial method* itself. The new requirement merely prescribes that the placebo effect must be taken into account, but it doesn't *specify* how exactly this should be done: it doesn't say anything about control groups or fake pills. In this sense, it is quite abstract. *The blind trial method* on the other hand is very specific, for it does specify what exactly must be done in order to void the possible placebo effect. It is easy to see that *the blind trial method* does not follow deductively from the conjunction of the placebo effect thesis and the requirement to accept only the best available theories. In particular, neither the placebo effect thesis nor the requirement to accept only the best available theories say anything about the necessity of giving fake pills to the control group. The only thing that can be deduced from the conjunction of the two is an abstract requirement that the possible placebo effect must be taken into consideration when assessing a drug's efficacy. The conjunction of the two does not specify how exactly this abstract requirement is to be implemented and, therefore, does not yield *the blind trial method*. In short, it is one thing to say that a new drug must be assessed in such a way that the placebo effect is taken into account and it is another thing to say that a new drug must be tested in a blind trial. There is an obvious logical gap between the two, for the new abstract requirement alone doesn't logically imply the blind trial method:

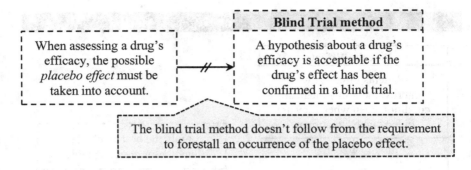

But, in this case, how come we ended up employing *the blind trial method*? In order to answer this question we must understand the nature of the relation between the requirement to forestall an occurrence of the placebo effect and the blind trial method.

It is readily seen that the blind trial method is a specification of the new requirement. While the new requirement is abstract ("the possible *placebo effect* must be taken into account"), the blind trial method is concrete, for it prescribes how exactly the testing should be done. Thus, the blind trial method specifies the new abstract requirement. This is the relation of *implementation*: a more concrete method implements the requirements of a more abstract method by making them more concrete. In our case, the blind trial method specifies the abstract new requirement by prescribing that a new drug must be tested in a version of a controlled trial where the active group receives real pills while the control group is given fake pills. In future diagrams, the relation of *implementation* will be depicted by an uncolored dashed arrow[29]:

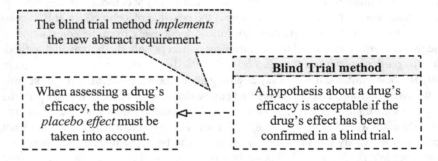

To understand the relation of implementation, we must appreciate that *the blind trial method* is by no means the only way of specifying the abstract requirement to account for the placebo effect. In principle, this abstract requirement can be specified in an infinite number of ways. For instance, instead of giving fake pills to the

---

[29] In object-oriented analysis, a similar relation holds between an *interface* and a *class* that realizes (implements) that interface. This explains my choice of an uncolored dashed arrow as a symbol for this relation.

patients in the control group we can make sure that the patients in the active group believe that they are not undergoing treatment. We can do this by giving them the drug in a concealed form (with food, drink etc.). Alternatively, if we had a nanotechnology capable of tracing every molecule of the drug in a patient's organism, we could establish whether the improvement had anything to do with the drug itself or whether it was due to the placebo effect more directly by tracing the causal interactions at a molecular level. In this hypothetical case, no fake pills or control groups would be necessary. With a bit of imagination one can think of many different implementations of the abstract requirement:

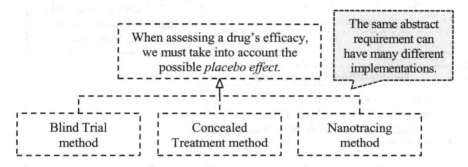

The relation of implementation is also apparent in the case of *the double-blind trial method*. If we leave the history of the case and consider it from the position of contemporary science, we can say that our knowledge about unaccounted effect, placebo effect, and experimenter's bias jointly produce additional abstract requirements:

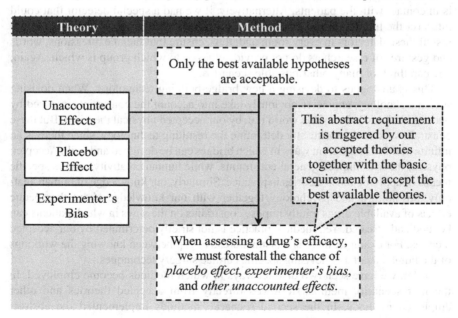

How exactly this abstract requirement can be specified is the next question. Our contemporary answer is *the double-blind trial method*. We employ the method because we believe that it forestalls the occurrence of unaccounted effects, placebo effect, and experimenter's bias. In other words, we employ it because we take it is an *implementation* of the abstract requirements which follow from our accepted theories:

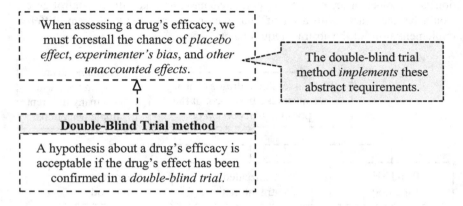

Yet, logically speaking, the double-blind trial method is not the only possible implementation of these abstract requirements. With a good deal of imagination one could invent many different ways in which these requirements could be specified. For one, instead of hiding the information from the researchers we could employ specially designed robot-researchers and, thus, make sure that no human-researcher is in contact with the patients. Alternatively, if we had a special detector that could interpret the hidden meanings of facial expressions, words and gestures, we could install these detectors in every room and, thus, ensure that facial expressions, words and gestures of researchers do not provide clues as to which group is which. Again, one can think of many other possible scenarios.

This is analogous to devising a new bridge-building technology. When devising such a technology, engineers obviously take into account the constraints imposed by our current knowledge of the world (i.e. by our accepted physical theories). But these constraints alone do not strictly determine the resulting technology, since there is an infinite number of different ways in which bridges can be designed and built. Accepted physical theories impose general constraints, while human creativity invents specific means of applying those theories in practice. Similarly, our knowledge of human anatomy, physiology, and psychology together with our knowledge about therapeutic effects of available drugs jointly impose constraints on the ways in which patients can be medically treated. Yet, medical practice is not strictly determined by our accepted theories. For example, there is a distinct creative gap between knowing the workings of the human heart and devising specific cardiac surgery techniques.

So far, we have seen two distinct ways in which methods become employed. In the first scenario, methods followed directly from accepted theories and other employed methods. In the second scenario, methods implemented the abstract requirements of other methods. Thus:

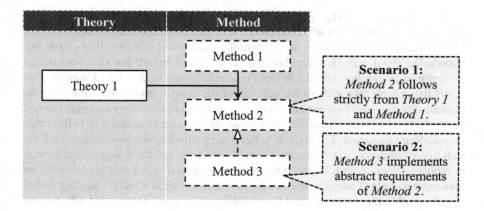

In the *first* scenario, the requirements of the method are strictly determined by the accepted theories and other employed methods (such as the fundamental requirement to accept only the best available theories). This is not the case in the *second* scenario, where methods are not strictly determined by the accepted theories and other employed methods. It takes a fair amount of ingenuity to devise concrete procedures that implement abstract requirements.

It can be shown, however, that these two scenarios are not totally disconnected, that there is one important similarity between the two. Consider *the double-blind trial method* that implements the requirement that the placebo effect, experimenter's bias, and other effects must be dealt with. Now, it is obvious that the method is based on our belief that by performing a double-blind trial we forestall the chance of unaccounted effects, placebo effect, and experimenter's bias:

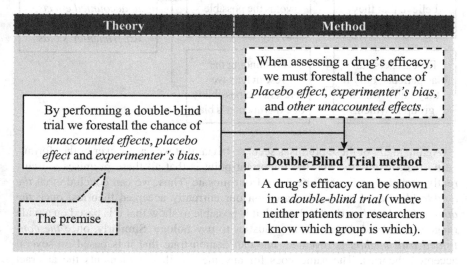

This premise, in turn, is based on several other propositions which can be deduced from a number of accepted theories. In particular, our belief that a trial with two similar groups minimizes the chance of unaccounted effects follows from our knowledge about statistical regularities, i.e. from our belief that two statistically similar groups can be expected to behave similarly *ceteris paribus*. Likewise, our belief that giving fake pills to the patients of the control group voids the possible placebo effect is based on our knowledge about human physiology and psychology, particularly on our tacit conviction that those who receive placebos will believe that they undergo treatment. Finally, our belief that by blinding the researchers we forestall the possibility of experimenter's bias follows from our knowledge about human psychology; namely, it follows from a seemingly trivial belief that researchers can bias patients only if they themselves know which group is which. Thus, our belief that the double-blind trial method forestalls the chance of unaccounted effects, placebo effect, and experimenter's bias is based on a number of accepted theories. Here is a detailed deduction:

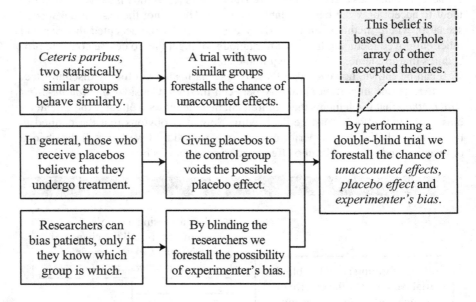

Again, many of these premises seem so trivial that we hardly ever bother formulating them explicitly. Yet, the important point is that all these premises are currently accepted and constitute part of our mosaic. Thus, we can conclude that *the double-blind trial method* is based on our currently accepted theories: once *the double-blind trial method* is devised, it is possible to show that it is based on a wide range of accepted theories from statistics to psychology. Similarly, once *the controlled trial method* is invented, one can demonstrate that it is based on several accepted theories. The same goes for any method that implements the abstract requirements of other methods.

In this sense, the two scenarios of method employment are similar: in both cases methods become employed only when they follow deductively from other elements of the mosaic. This brings us to *the third law*: a method becomes employed only when it is deducible from other employed methods and accepted theories of the time. Let us give its extended formulation:

> **3rd Law: Method Employment (Extended)**
>
> A method becomes employed only when it is deducible from other employed methods and accepted theories of the time, i.e. either (1) when it strictly follows from the other employed methods and accepted theories, or (2) when it implements some abstract requirements of other employed methods.

The cases illustrating *the third law* that I have discussed so far concerned exclusively the history of drug testing methods. HSC provides us with many other illustrations of the mechanism of method employment.

One example is the Aristotelian method employed in natural philosophy up until the end of the seventeenth century. It can be shown that it implemented the constraints imposed by the theories accepted at the time. One of the cornerstones of the Aristotelian natural philosophy was the belief that everything has its *nature*, an essential quality that makes a thing what it is, i.e. a quality without which a thing ceases to be what it is. The nature of, say, an acorn is that it is a potential oak tree, whereas the nature of a man is his capacity of reason. It followed from this, that knowing a thing amounts to knowing its nature, i.e. that the best theory is the one that uncovers the nature of the thing. From this premise and the basic requirement to accept only the best available theories it follows that a theory is acceptable only if it grasps the nature of the thing under study:

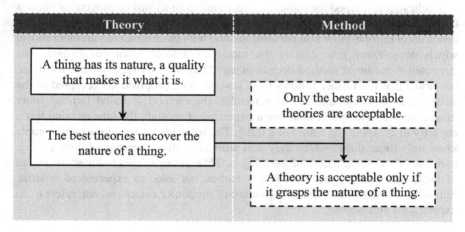

But this requirement is too abstract – it may be specified in a number of different ways. Before I turn to the implementation of this requirement by *the Aristotelian-*

medieval method, I must repeat that explicating any actually employed method is not an easy task, especially when it is not the only possible implementation of the abstract requirements. Thus, as with any reconstruction of any other employed method, my reconstruction is just another historical hypothesis which is not immune to revision.

Now, if my explication of the Aristotelian-medieval method is correct, then essentially it required that a scientific theory should be an axiomatic-deductive system of propositions where the axioms grasp the nature of a thing under study and theorems follow deductively from the axioms. The axioms were expected to be *intuitive* in the sense that any person experienced enough in the subject should be able to appreciate them. For instance, so the story goes, one who is experienced with mankind will undoubtedly see that the nature of man, his most essential quality that differentiates him from other animals, is his capacity of reason. Thus, in the Aristotelian-medieval mosaic, the proposition "man is a thinking animal" is one of the axioms from which many theorems can be deduced. If we take this axiom in conjunction with another fundamental principle of Aristotle's philosophy that every natural thing tends towards the fulfilment of its nature (e.g. an acorn tends towards becoming an oak tree), we can deduce that man's function consists in exercising his capacity of reason to the fullest. In short, we can word the requirements of the Aristotelian-medieval method thus:

Aristotelian-Medieval method

A proposition is acceptable if it grasps the
nature of a thing through intuition
schooled by experience, or if it is deduced
from general intuitive propositions.

I believe this method (or something very similar to it) was employed for about a millennium – for about five centuries in Arab-Islamic science and then another five centuries in Europe. Now, compare this method with the abstract requirement above, which doesn't say how exactly the nature of a thing must be grasped. The Aristotelian-medieval method does *implement* that abstract requirement, for it can be shown that the former follows from several other propositions accepted in the Aristotelian-medieval mosaic. In particular, the method is based (among other things) on the assumption that, when experienced enough, the human mind has a capacity of grasping the nature of a thing. The basic idea is that, in general, people know only those things which they deal with and, when experienced with a thing, they can tell what the nature of the thing is. Thus, if one wishes to know the nature of bees, one doesn't go to his fellow barber, but asks an experienced apiarist. Similarly, if one is interested in the nature of motion or causation, one refers to an experienced philosopher.

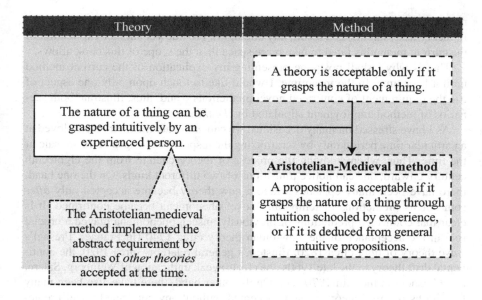

It is obvious that the Aristotelian-medieval method is not the *only* possible way in which this abstract requirement of grasping the nature of a thing can be implemented – it is just *one* of the possible ways. The general scheme of this case is identical to that of the placebo effect case – accepted theories impose some abstract requirements, which are specified by means of other accepted theories.

One further illustration of the third law in action is provided by the history of the employment of *the hypothetico-deductive method*. It must be stressed that what I call hypothetico-deductive *method* should be kept apart from the so-called hypothetico-deductive *methodology*. For lack of a better term, I use the same label to denote the method that has been actually employed in natural philosophy and then natural science from circa the early eighteenth century up until now. If explicating the Aristotelian-medieval method was a tricky task, the task of explicating the current method of natural science is tenfold more difficult. Indeed, what are the implicit expectations of our contemporary scientific community? Should a theory necessarily provide confirmed novel predictions in order to become accepted or is it possible for a theory to become accepted without any confirmed novel predictions? And if novel predictions are mandatory, how should *novelty* be understood (in a temporal sense, in a heuristic sense, or in some other sense)?[30] Also, should a new theory somehow cover all the phenomena accounted for by the currently accepted theory, or can it become accepted even if it leaves some known facts unexplained? In order to provide a thorough explication of the current method of natural science we would have to address many similar questions. What complicates the situation further is the fact that several generations of philosophers have attempted to portray their *methodologies* as the correct explication of the method of science. The likes of Whewell, Mill, Popper, Lakatos, and the early Laudan all believed that the require-

---

[30] For references, see footnote 5 on page 101.

ments of their own methodologies are actually employed by the scientific community in theory assessment. As a result, any attempt to provide yet another explication of the current method is fated to take more space than the scope of this book allows.

That is why I shall postpone presenting my explication of the current method until a later time. In the meantime, I would like to touch upon only one aspect of it – the requirement of confirmed novel predictions – and, thus, illustrate both scenarios of method employment stipulated by *the third law*.[31]

As I have stressed on many occasions, we can explicate the method employed at a particular time period only by scrutinizing the respective changes in the mosaic of the time. Thus, when studying the process of theory change from the eighteenth century onward, one can notice transitions of two different kinds. On the one hand, there have been many episodes where a new theory became accepted only *after* some of its predictions of temporally novel phenomena became confirmed. It is clear that confirmed predictions of previously unseen phenomena played a crucial role in the acceptance of the Newtonian theory circa 1740 (in France), Fresnel's wave theory of light circa 1820, Einstein's general relativity circa 1920, the continental drift theory in the late 1960s, the electroweak unification of Weinberg, Salam, and Glashow in the mid-1970s etc. On the other hand, there have also been many transitions where a theory became accepted without any confirmed novel predictions whatsoever.[32] It is easy to see that confirmed novel predictions played no role in the acceptance of Mayer's lunar theory in the 1760s, Coulomb's inverse square law in the early 1800s, the three laws of phenomenological thermodynamics in the 1850s, Clark's formulation of the law of diminishing returns in economics circa 1900, quantum mechanics circa 1927 etc.[33]

This, I believe, indicates that we do expect confirmed novel predictions but only in very special circumstances. Interestingly, there was one common characteristic in all those episodes when a theory's acceptance required confirmed novel predictions. What all those theories had in common is that they all altered our views on the structural elements of the world by positing the existence of absolute space and time (Newton), waves of light (Fresnel), a curved spacetime continuum (Einstein), neutral current interactions (electroweak theory) etc. In other words, all these theories introduced modifications to what can be roughly called *accepted ontology* (or, to use Godfrey-Smith's terminology, "hidden structure of the world"[34]). Apparently, we do require confirmed novel predictions only if a theory attempts to introduce changes in the accepted view on the structural elements of the world, i.e. if it

---

[31] Again, it should be borne in mind that what follows is merely another *historical hypothesis*. I cannot vouch that the future HSC won't find it faulty – that is not my thesis here. My thesis is that the mechanism of method employment expressed in *the third law* is correct.

[32] It is not surprising, therefore, that several generations of philosophers have debated over the role of novel predictions, with the likes of Whewell, Popper and Lakatos on one side and the likes of Mill, Keynes and Laudan on the other. See footnote 5 on page 101 for references.

[33] Note that the dates of acceptance are all tentative. As I have indicated in section "Relevant Facts", many contemporary narratives focus on theory *construction*, rather than *acceptance*. As a result, establishing the exact date or even the decade when a theory became accepted is often quite a challenging task.

[34] See Godfrey-Smith (2003), pp. 211–212.

introduces new entities, new interactions, new processes, new particles, new waves, new forces, new substances etc.

In contrast, if a theory does *not* attempt to alter our views about the constituents of the world, we do not require any confirmed novel predictions – it can become accepted even in the absence of novel predictions. For instance, Mayer's lunar theory wasn't expected to have confirmed novel predictions since it didn't attempt to change anything in the accepted ontology; it was merely attempting to come up with a proper model of lunar motion that would allow calculating the future positions of the moon with necessary accuracy and precision. As a result, all that the scientific community seemed to be interested in was whether the theory fit the observational data with the desired level of accuracy and precision. Similarly, Coulomb's law was an attempt to quantify a particular physical relationship and, naturally, the community didn't expect it to predict any previously unseen phenomena.

In short, if you present an equation that sums up a relation between various quantities, you don't need novel predictions. Similarly, if you attempt to explain a certain phenomenon by employing only the elements of the currently accepted ontology, you don't need novel predictions. But if you are trying to convince the community that there exists some new *type* of particle, substance, interaction, process, force, etc., you *must* provide confirmed novel predictions. After all, if someone tries to persuade us that space is not Euclidean and that there is in fact a curved spacetime continuum, it will take nothing short of actual light-bending to convince us. Similarly, if a biological theory posits that species are mutable and that they are products of the process of evolution, we won't be convinced unless we discover fossils of extinct species. Likewise, if a theory attempts to modify the accepted ontology by positing the existence of superstrings, there is no way of convincing the scientific community other than by confirmed novel predictions. Thus, what we nowadays seem to expect can be explicated along these lines:

Confirmed Novel Predictions requirement

If a theory attempts to alter the accepted views
on the structural elements of the world (the
accepted ontology), it must have confirmed
novel predictions in order to become accepted.

It is my historical hypothesis that this requirement has been one of the key ingredients of the method actually employed in natural science since the eighteenth century.[35] Let us assume for the sake of argument that this historical hypothesis is correct. *The third law* stipulates that the requirement of confirmed novel predictions could become employed only if it was a deductive consequence of the accepted theories and other employed methods of the time. So a question arises: what theories

---

[35] It is readily seen that my explication of this requirement is in tune with Carl Sagan's intuition that "extraordinary claims require extraordinary evidence". See Sagan (1980), 01:24.

and methods does this requirement follow from? I think it is based on at least two fundamental assumptions accepted since the eighteenth century.

One fundamental principle that has been accepted since the early eighteenth century is that the world is more complex than it appears in observations, that there is more to the world than meets the eye.[36] From this assumption, it follows that what appears in observations may be an effect of some inner structure which is not directly observable. It is safe to say that this proposition is still implicit in our contemporary mosaic, for we still take it for granted that all phenomena are produced by some more fundamental structures and processes. That is precisely the reason why we tolerate hypotheses concerning unobservable entities, forces, fields, etc.

In addition, it has been accepted since the early eighteenth century that, in principle, any phenomenon can be produced by an infinite number of different underlying mechanisms. For instance, two clocks, which look absolutely identical from the outside, may nevertheless have two completely different arrangements of cogwheels, springs, and levers. This leads us to the thesis of underdetermination that, in principle, any finite body of evidence can be explained in an infinite number of ways.[37] Recall how many different explanations the phenomenon of falling stone has received. While Aristotle believed the stone was a heavy object descending towards the centre of the universe, Descartes believed the stone was being displaced by a greater centrifugal force of faster particles in the Earth's vortex. While for Newton, it was accelerated motion in the gravitational field of the Earth, for Einstein it was inertial motion in a curved spacetime continuum. But if any data can be shoehorned into an infinite number of different theories, it is quite easy to end up accepting some cooked-up explanation. The risk is especially high when an explanation is provided *post hoc* (after the fact). Indeed, it is always possible to hypothesize a mechanism which reproduces the known data with utmost accuracy and precision but nevertheless is totally incorrect. The classic example is provided by the early-modern version of the Ptolemaic astronomy with its constantly growing number of epicycles which nevertheless provided highly accurate predictions.[38]

The abstract requirement that follows from these two principles is that whenever we assess a theory that introduces some new internal mechanisms (new types of substances, particles, forces, fields, interaction, processes etc.) we must take into account that this hypothesized internal mechanism may turn out to be fictitious even if it manages to predict the known phenomena with utmost precision. In other words, we do not tolerate "fiddling" with the *accepted ontology*; if a theory attempts to modify the

---

[36] This idea was expressed in many different ways by many authors. It is implicit in Locke's distinction between primary and secondary qualities, in Newton's belief that the rays of light that produce our sensation of color are, themselves, not colored, and in Descartes's belief that the only attribute of matter is extension and that such qualities as taste, smell or color are effects produced upon our senses by the configuration and motion of material particles.

[37] This thesis must not be confused with a stronger thesis held be Quine. See Laudan (1996), pp. 29–54 for discussion.

[38] There is a humorous YouTube video called "Ptolemy and Homer Simpson" that illustrates the point.

accepted ontology, it must show that it is not cooked-up. Note that the employment of this abstract requirement illustrates the first scenario of method employment:

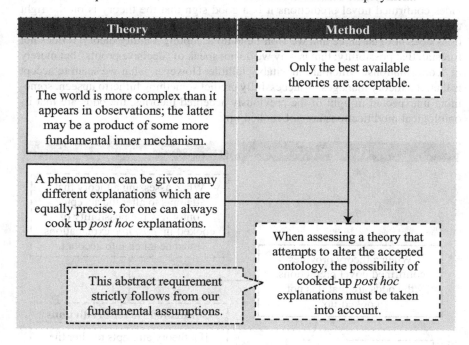

This requirement is abstract, for it does not specify how exactly we must filter out cooked-up explanations. One possible implementation of this requirement is the above discussed requirement of confirmed novel predictions:

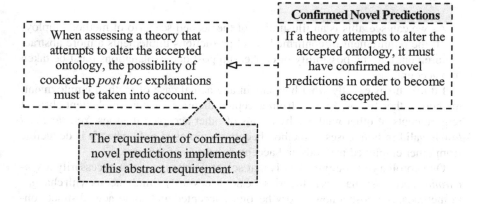

This requirement of confirmed novel predictions is itself based on another assumption, tacitly accepted by the scientific community that when a theory provides confirmed novel predictions it is a good sign that the theory is on the right track. Now, we clearly understand that the confirmation of a theory by novel predictions does not guarantee that we won't end up accepting a faulty theory. It does not forestall that possibility; that is why we do not speak of "decisive proofs" but merely of "confirmations" which are inevitably fallible. However, what we seem to accept is that, if a theory manages to successfully predict something hitherto unseen, something unexpected in light of the previously accepted theories, the odds are that its ontological modifications are not cooked-up:

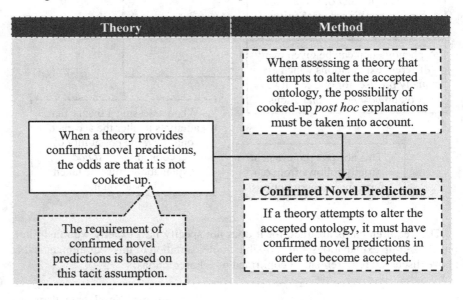

As we can see, this is an illustration of the second scenario of method employment: the requirement of confirmed novel predictions implements a more abstract requirement that the possibility of cooked-up post hoc explanations must be taken into account.

I think this is exactly what happens in any actual case of method employment: methods either follow directly from accepted theories, or they implement abstract requirements of other methods by means of other accepted theories. Yet, *the third law* is valid in both cases: a method becomes employed only when it is deducible from other employed methods and accepted theories of the time.

One corollary of the third law is that scientific change is not necessarily a *synchronous* process: changes in theories are not necessarily simultaneous with changes in methods. Suppose a new theory becomes accepted and some new abstract constraints become imposed. In this case, we can say that the acceptance of a theory resulted in the employment of a new method and the employment of a new method was synchronous with the acceptance of a new theory. But we also know that there is the second scenario of method employment, where a method implements some abstract requirements of other employed methods. In this scenario, there is a certain creative gap between abstract requirements that follow directly from accepted theories and methods that implement these abstract requirements. Devising a new method that would implement abstract requirements takes a fair amount of ingenuity and, therefore, there are no guarantees that these abstract requirements will be immediately followed by a new concrete method. In short, changes in methods are not necessarily simultaneous with changes in theories:

3rd Law: Method Employment	Asynchronism of Method Employment
A method becomes employed only when it is deducible from other employed methods and accepted theories of the time.	The employment of new methods *can be* but is not *necessarily* a result of the acceptance of new theories.

This corollary is in obvious contradiction with Kuhn's belief that scientific change is essentially a *wholesale* process, where any transition from one paradigm to another involves changes in theories and methods alike. In Kuhn's view, a paradigm is an inextricable mix of theories and methods and, thus, changes in theories are necessarily synchronous with changes in methods. For example, a transition to the Newtonian paradigm is, according to Kuhn, a transition both to the Newtonian theory (including the ontology that comes with it) and to the Newtonian rules of method. Scientific change, in Kuhn's view, is necessarily a synchronous process.[39]

The third law makes clear why this is not the case. While it is true that new theories impose new methods, it is not true that the employment of methods is *always* a result of changes in theories. It could take years or even centuries for an abstract method to become implemented by a concrete method. Consider the following example.

A long time ago, we came to realize that, when it comes to acquiring data about such minute objects as molecules or living cells, the unaided human eye is virtually useless. This proposition yields, among other things, an abstract requirement that, when counting the number of cells, the resulting value is acceptable only if it is obtained with an "aided" eye. This abstract requirement has been implemented in a variety of different ways. First, there is *the counting chamber method* where the cells are placed in a counting chamber – a microscope slide with a special sink – and the number of cells is counted manually under a microscope. There is also *the plating method* where the cells are distributed on a plate with a growth medium and each cell gives rise to a single colony. The number of cells is then deduced from the number of colonies. In addition, there is *the flow cytometry method* where the cells are hit by a laser beam one by one and the number of cells is counted by means of detecting the light reflected by the cells. Finally, there is *the spectrophotometry*

---

[39] Laudan, who criticizes Kuhn's wholesale change view, illustrates his position by several historical examples. Unfortunately, all of his examples apply to changes in *methodologies*, rather than actually employed *methods*. See Laudan (1984), pp. 69–73, 78, 80–84, 95. See also section "Explicit and Implicit" above for discussion.

method where the number of cells is obtained by means of measuring the turbidity in a spectrophotometer. Now, each of these methods implements the abstract requirement that the number of cells should be obtained only by an "aided" eye[40]:

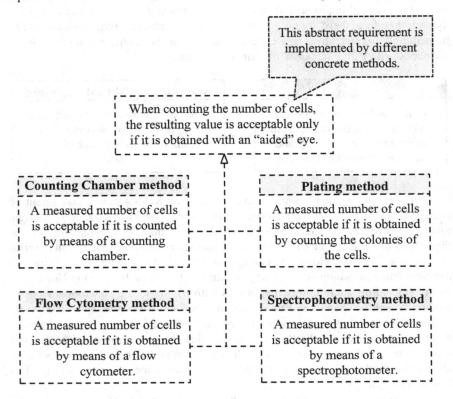

What is important in the current context is that these implementations were devised and became employed at different times. For instance, if *the counting chamber method* became employed as early as circa 1880s, *the flow cytometry method* became employed only in the 1950–60s. This is exactly the idea expressed in *the asynchronism of method employment* corollary.

The third law has many other interesting consequences. However, before we move on to discussing them, we must complete our discussion of the laws.

The Zeroth Law: Compatibility

The first question that is likely to arise after a quick glance at our final law is why such a strange numbering. Why "the zeroth" and not "the fourth"? The reason for this strange numbering is simple: when we consider all four laws, it is immediately

---

[40] Note that this list of methods is not *exhaustive*: there are other implementations as well. In addition, new implementations of old requirements are also always possible.

noticeable that only one of them – *the zeroth law* – applies to the mosaic viewed from a *static* perspective, while the other three apply to the mosaic viewed from a *dynamic* perspective (i.e. they concern *changes* in the mosaic). In more technical terms, only the zeroth law is synchronic whereas the other three are diachronic. The zeroth law stipulates that if we could "pause" the process of scientific change we would notice that its elements are mutually compatible. Thus, *the zeroth law* can also be called *the law of compatibility*. Compatibility of the elements is one of very few things that can be said about the mosaic if it is considered from a static perspective. This is why I have chosen "zero" as the number for *the law of compatibility*:

> **0th Law: Compatibility**
>
> At any moment of time, the elements of the
> scientific mosaic are compatible with each other.

A short historical note is in order here. In its initial formulation (proposed in 2012), the zeroth law stated that at any moment of time, the elements of the mosaic are *consistent* with each other, where consistency was understood in a classical logical sense. However, in 2013, it was shown by Rory Harder that the original formulation of the zeroth law is untenable for both logical and historical reasons. For one, while ascertaining that no two accepted propositions are mutually inconsistent might be a viable task in a mosaic with only a handful of propositions, the task may prove virtually impossible in a more complex mosaic. In addition, even if the community somehow manages to ascertain logical consistency of all *openly* accepted propositions, there will still remain a possibility that some of the *logical consequences* of two accepted theories are mutually inconsistent. Obviously, no scientific community is in a position to trace all the logical consequences of all accepted theories and, therefore, there is always a chance that accepted *Theory 1* and accepted *Theory 2* can have contradictory logical consequences:

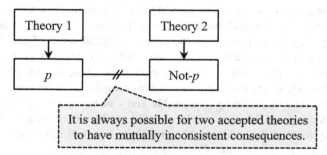

However, the main point is that, in principle, it is possible for a scientific community to *knowingly* accept a contradiction. There might be cases when the cost of rejecting two mutually inconsistent theories is greater than that of their simultaneous acceptance. It may so happen that *Theory 1* and *Theory 2* are the best available descriptions of their respective domains and yet the two theories are mutually inconsistent. In such cases, the community may opt for keeping both theories in the mosaic despite their mutual inconsistency. In fact there have been several historical cases when the community accepted two logically inconsistent theories as the best available descriptions of their respective domains. The conflict between general relativity and quantum physics is probably the most famous illustration of this phenomenon. We normally take general relativity as the best description of the world at the level of massive objects and quantum physics[41] as the best available description of the micro-world. But we also know that, from the classical logical perspective, the two theories contradict each other. The inconsistency of their conjunction becomes apparent when they are applied to objects that are both extremely massive and extremely small (i.e. a singularity inside a black hole). Yet, despite the existence of this contradiction, the community accepts both theories as the best available descriptions of their respective domains.

Consequently, the original formulation of *the zeroth law* was deemed untenable for both historical and theoretical reasons. *The zeroth law* was modified by Rory Harder to take into account the fact that mutual compatibility of two theories is not necessarily decided on the basis of their logical consistency. In its current formulation, *the zeroth law* stipulates that elements of the mosaic are mutually compatible, while compatibility or incompatibility of the elements is decided by the criteria of compatibility employed in the mosaic.

It is therefore crucial not to confuse the notion of *compatibility* with the concept of *consistency* of classical logic. The formal definition of inconsistency is that a set is inconsistent just in case it entails some sentence and its negation, i.e. p and not-p. The classical logical principle of noncontradiction stipulates that p and not-p cannot be true. In classical logic, the major problem with accepting a contradiction is that p and not-p imply every other sentence. In other words, classical logic is said to be explosive for a contradiction entails triviality. The classical logical principle of noncontradiction helps to avoid such trivialities by stipulating that two mutually contradicting propositions cannot be both true. In contrast, the notion of compatibility implicit in *the zeroth law* is much more flexible, for its actual content depends on the criteria of compatibility employed at a given time. As a result, the actually employed criteria of compatibility can differ from mosaic to mosaic. While in some mosaics compatibility may be understood in the classical logical sense of consistency, in other mosaics it may be more flexible and allow for two contradictory theories to be simultaneously accepted under certain circumstances. Thus, in principle, there can exist such mosaics, where two theories that are inconsistent in the classical logical sense are nevertheless mutually compatible and can be simultaneously accepted within the same mosaic. In other words, a mosaic can be inconsistency-intolerant or inconsistency-tolerant depending on the criteria of compatibility employed by the scientific community of the time. Consider two hypothetical cases.

First, imagine a community that believes that all of their accepted theories are absolutely (demonstratively) true. This *infallibilist* community also knows that,

---

[41] More precisely: *some* of its propositions. Refer to section "Time, Fields, and Scale" for details.

according to classical logic, p and not-p cannot be both true. Since, according to this community, all accepted theories are strictly true, the only way the community can avoid triviality is by stipulating that any two accepted theories must be mutually consistent. In other words, *by the third law*, they end up employing the classical logical law of noncontradiction as their criterion of compatibility:

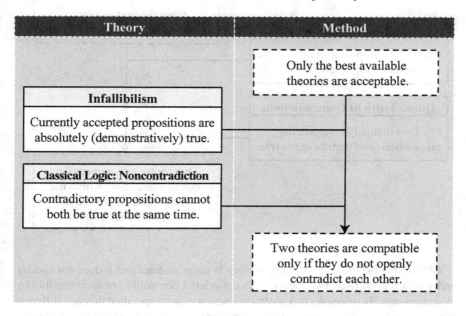

Now, imagine another community that accepts the position of *fallibilism*. This community holds that no theory in empirical science can be demonstratively true and, consequently, all accepted empirical theories are merely quasi-true (i.e., approximately true, truthlike). But if any accepted empirical theory is only quasi-true, it is possible for two accepted empirical theories to be mutually inconsistent. In other words, this community accepts that two contradictory propositions may both contain grains of truth, i.e. to be quasi-true.[42] In order to avoid triviality, this community employs a paraconsistent logic, i.e. a logic where a contradiction does not imply everything.[43] This fallibilist community does not necessarily reject classical logic; it merely realizes that the application of classical logic to quasi-true propositions entails triviality. Thus, the community also realizes that the application of classical principle of noncontradiction to empirical science is problematic, for no empirical theory is strictly true. As a result, by *the third law*, this community employs criteria of compatibility very different from those employed by the infal-

---

[42] See Bueno et al. (1998).

[43] It can be also noted that the *criteria of deductive acceptance* employed by this fallibilist community are based on the laws of inference of a paraconsistent logic, which are much weaker than the classical laws of inference. See the discussion in section "Static and Dynamic Methods" below.

libilist community; two theories are no longer expected to be mutually consistent in
order to be considered compatible:

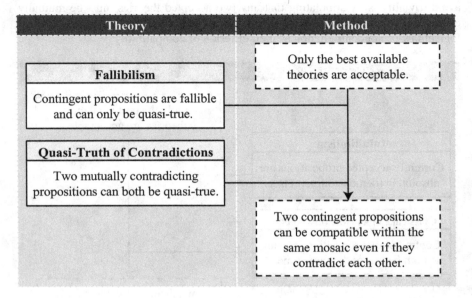

Obviously, this criterion of compatibility is quite abstract and it does not specify
under which conditions two theories are considered compatible or incompatible by
the community. In principle, this abstract criterion may be specified in many different
ways. For instance, the community may consider two inconsistent theories compatible
as long as they can be limited to two different domains (e.g. a physical theory and a
chemical theory may be mutually inconsistent yet compatible in the same mosaic):

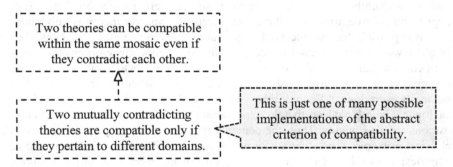

Here we need not delve into possible implementations of this abstract criterion.
Suffice it to appreciate that a community with non-classical criteria of compatibility
is conceivable.[44] This is essentially the main idea of the zeroth law. In its current

---

[44] It is not clear whether contradictions can be accepted only by fallibilist communities, i.e. whether
an infallibilist community can also knowingly accept contradictions. According to Rory Harder
such a scenario is possible, while I am inclined to think that contradictions can only be openly
accepted by fallibilist communities. This is an open question and calls for a further clarification.

version, the zeroth law simply states that the compatibility or incompatibility of any two elements is decided by the criteria of compatibility employed at the time. Whether a given mosaic is or isn't inconsistency-tolerant depends on the criteria of compatibility employed in the mosaic.

Although many authors have expressed the idea of the mutual agreement of elements, it is safe to say that the historical character of compatibility criteria has not been properly appreciated. Thus, in Otto Neurath's conception, the idea of mutual agreement of the elements played a central role. Neurath compared scientists with sailors who rebuild their ship on the sea by replacing one beam at a time. A new beam, on this view, is added only when it fits the rest of the ship.[45] It is even safe to say that, in Neurath's conception, mutual agreement of the elements is basically the only guiding principle of scientific change. Similarly, in Quine's view, we adjust and replace the elements of the so-called *web of belief* by maintaining the mutual agreement between the elements.[46] Similar views are implicit in a vast majority of conceptions of scientific change. Yet, it has been often tacitly assumed that compatibility of any two elements is decided by the law of noncontradiction of classical logic: if propositions are mutually inconsistent, they are incompatible. Effectively, the possibly changeable character of compatibility criteria and the mechanism of their employment has not been properly understood prior to the reformulation of *the zeroth law*.[47] In contrast, *the zeroth law* emphasizes that two theories, which are considered compatible by the criteria of one mosaic, may turn out to be incompatible by the criteria of another mosaic.

The law of compatibility has three closely linked aspects. First, it states that two *theories* simultaneously accepted in the same mosaic cannot be incompatible with one another. It also states that at any moment two simultaneously employed *methods* cannot be incompatible with each other. Finally, it states that, at any moment of time, there can be no incompatibility between accepted *theories* and employed *methods*. Let us consider these three aspects of the law of compatibility in turn.

It must be emphasized that, when applied to *theories*, *the law of compatibility* only covers theories *accepted* in the same mosaic. It says nothing about *pursued* or *used* theories.[48] While two mutually incompatible theories cannot be simultaneously *accepted*, it is obvious that they can be simultaneously *pursued* or simultaneously *used*.

As I have explained in section "Acceptance, Use, and Pursuit", when we search for a solution to a given technical problem, we often draw on theories that cannot be simultaneously accepted. Mutually incompatible theories are often simultaneously *used* even in the same project. For instance, circa 1600 astronomers could easily use both Ptolemaic and Copernican astronomical theories to calculate the ephemerides

---

[45] See Neurath (1973), p. 199.

[46] See Quine and Ullian (1978).

[47] Shapere comes close to the new formulation of *the zeroth law* when he says that the requirement of consistency of our theories is in principle revisable. Yet, Shapere doesn't explain what makes two theories compatible in one mosaic and incompatible in another. See Shapere (1980), pp. 235–237.

[48] See section "Acceptance, Use, and Pursuit" above.

of different planets. Similarly, in order to obtain a useful tool for calculating atomic spectra, Bohr mixed some propositions of classical electrodynamics with a number of quantum hypotheses.[49] Finally, when nowadays we build a particle accelerator, we use both classical and quantum physics in our calculations. Thus, sometimes propositions from two or more incompatible theories are mixed in order to obtain something practically useful.

The same goes for the *pursuit* of incompatible theories. There is nothing extraordinary in the fact that mutually incompatible theories can be simultaneously pursued. After all, scientific change wouldn't be possible if we didn't pursue different alternatives. Moreover, even an individual alternative can contain incompatible propositions. Take for instance, Clausius's attempt to derive Carnot's theorem, where he employed two incompatible theories of heat – Carnot's *caloric theory of heat*, where heat was considered a fluid, and also Joule's *kinetic theory of heat*, where the latter was conceived as a "force" that can be converted into work.[50] Thus, the existence of incompatible propositions in the context of pursuit is quite obvious. There is good reason to believe that "reasoning from an inconsistent theory usually plays an important heuristic role"[51] and that "the use of inconsistent representations of the world as heuristic guideposts to consistent theories is an important part of scientific discovery".[52] Yet, we must keep in mind that this has nothing to do with the mosaic of *accepted* theories, just as *the law of compatibility* has nothing to say about either *use* or *pursuit*. The law of compatibility applies only to the mosaic of *accepted* theories.

Naturally, it assumes that the community employs certain compatibility criteria. As with any criteria, explication of the compatibility criteria of a given time period is not a simple task and calls for a special historical study. Here I can only suggest one vague hypothesis which is based on my study of a very limited number of cases. First, we need to appreciate that the contemporary community appears to be employing different criteria of compatibility for theories in different fields of inquiry. While in formal science (logic, mathematics) we seem to be inconsistency-intolerant, in empirical science we are prepared to be more flexible and tolerate formal logical inconsistencies under certain circumstances. It is true, of course, that we are never happy about inconsistencies and we do our best to eliminate them wherever possible, but the point is that we are prepared to tolerate logically inconsistent propositions within our mosaic when certain conditions are met. I think there are at least two distinct scenarios when the contemporary community is prepared to tolerate formal logical inconsistencies in the mosaic.

In the first scenario, we seem to be prepared to accept two mutually inconsistent propositions into the mosaic provided that they do not have the same object. More specifically, two propositions seem to be considered compatible by the contemporary community when, by and large, they explain different phenomena, i.e. when

---

[49] For a discussion of the case, see Smith (1988).

[50] The case is scrutinized in Meheus (2003).

[51] Meheus (2003), p. 131.

[52] Smith (1988), p. 429.

they have sufficiently different fragments of reality as their respective objects. When determining the compatibility or incompatibility of any two theories, the community seems to be concerned with whether the theories can be limited to their specific domains. Suppose *Theory 1* provides descriptions for phenomena A, B, and C, while *Theory 2* provides descriptions for phenomena C, D, and E. Suppose also that the descriptions of phenomenon C provided by the two theories are inconsistent with each other. Thus:

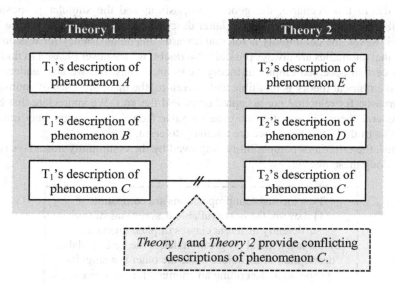

Theory 1 and *Theory 2* provide conflicting descriptions of phenomenon C.

Although the two theories are logically inconsistent, normally this is not an obstacle for the contemporary scientific community. Once the contradiction between the two theories becomes apparent, the community seem to be limiting the applicability of at least one of the two theories by saying that its laws do not apply to phenomenon C. While limiting the domains of applicability of conflicting theories, we may still believe that the laws of both theories should *ideally* be applicable to phenomenon C. Yet, we understand that *currently* their laws are not applicable to phenomenon C. In other words, we simply concede that our *current* knowledge of phenomenon C is deficient.

One textbook example of this point is our current stance on the apparent conflicts between general relativity and quantum physics. While we admit that *ideally* singularities within black holes must be subject to the laws of both theories, we also realize that *currently* the existing theories cannot be consistently applied to these objects, for combining the two theories is not a trivial task. Consequently, we admit that there are many aspects of the behaviour of these objects that we are yet to comprehend. Thus, it is safe to say that nowadays we accept the two theories only with a special "patch" that *temporarily* limits their applicability, while pursuing many different theories of quantum gravity.[53] Nowadays, the explicit statement of the

---

[53] For a short history of these developments, see Rovelli (2000).

known limitations of the two theories can be found in almost any textbook or encyclopaedia article on the topic.[54] It appears as though the reason why the community considers the two theories compatible despite their mutual inconsistency is that these theories are the best available descriptions of two considerably different domains.

In the second scenario, we are normally willing to tolerate inconsistencies between an accepted *general* theory and a *singular* proposition describing some anomaly. In this scenario, the general proposition and the singular proposition describe the same phenomenon; the latter describes a counterexample for the former. However, the community is tolerant towards this inconsistency for it is understood that anomalies are always possible. No doubt, we are never pleased to find out that a certain accepted empirical theory faces anomalies, but we also understand that no empirical theory is infallible and, therefore, the mere presence of anomalies is no reason for rejecting our accepted empirical theories. We appreciate that both the general theory in question and the singular factual proposition may contain grains of truth. In this sense, we are anomaly-tolerant.[55]

Thus, the criteria of compatibility employed by the community nowadays seem to be along these lines[56]:

> Two inconsistent propositions are compatible if
> (1) they are the best available descriptions of two
>   sufficiently different classes of phenomena, or
> (2) one is a general proposition that is the best available
>   description of its object while the other is a singular
>   proposition describing an anomaly for the former.

It can be argued that our contemporary criteria of compatibility have not always been employed. Consider the case of the reconciliation of the Aristotelian natural philosophy and metaphysics with Catholic theology. As soon as most works of Aristotle and its Muslim commentators were translated into Latin (circa 1200), it became obvious that some propositions of Aristotle's original system were inconsistent with several dogmas of the then-accepted Catholic theology. Take, for instance, the Aristotelian conceptions of *determinism, the eternity of the cosmos*, and *the mortality of the individual soul*. Evidently, these conceptions were in direct conflict with the accepted Catholic doctrines of *God's omnipotence* and *free will*, of *creation*, and of *the immortality of the individual human soul*.[57] Moreover, some of

---

[54] See, for example, Ghirardi (2005), pp. 344–357; Penrose (2004), pp. 849–853.

[55] A number of historical examples of anomaly-tolerance are discussed in section "The First Law: Scientific Inertia" above.

[56] Again, it is quite likely that my explication is incorrect; there is no doubt that it will be modified by professional historians. My only goal here is to give a preliminary, albeit vague, idea as to what the contemporary compatibility criteria require.

[57] For discussion, see Lindberg (2008), pp. 228–253.

the passages of Scripture, when taken literally, appeared to be in conflict with the propositions of the Aristotelian natural philosophy. In particular, Scripture seemed to imply that the Earth is flat (e.g. Daniel 4:10–11; Mathew 4:8; Revelation 7:1), which was in conflict with the Aristotelian view that the Earth is spherical. It is no surprise, therefore, that many of the propositions of the Aristotelian natural philosophy were condemned on several occasions during the thirteenth century.[58] To resolve the conflict, Albert the Great, Thomas Aquinas and others modified both the Aristotelian natural philosophy and the biblical descriptions of natural phenomena to make them consistent with each other. On the one hand, they stipulated that the laws of the Aristotelian natural philosophy describe the natural course of events only insofar as they do not limit God's omnipotence, for God can violate any laws if he so desires. Similarly, they modified Aristotle's determinism by adding that the future of the cosmos is determined by its present only insofar as it is not affected by free will or divine miracles. Similar modifications were introduced to many other Aristotelian propositions. On the other hand, it was also made clear that biblical descriptions of cosmological and physical phenomena are not to be taken literally, for Scripture often employs a simple language in order to be accessible to common folk. Thus, where possible, literal interpretations of Scripture were supposed to be replaced by interpretations based on the Aristotelian natural philosophy.[59] Importantly, it is only after this reconciliation that the modified Aristotelian-medieval natural philosophy became accepted by the community.[60]

This and similar examples seem to be suggesting that the compatibility criteria employed by the medieval scientific community were quite different from those employed nowadays. While apparently we are inconsistency-tolerant (at least when dealing with theories in empirical science), the medieval scientific community was inconsistency-intolerant in the sense that they wouldn't tolerate any open inconsistencies in the mosaic. Their criteria of compatibility were probably along these lines:

> Two theories are compatible if they do not openly contradict each other.

Once again, it needs to be emphasized that this whole reconstruction is just another historical hypothesis which I present here for illustrative purposes; it must be further scrutinized by professional historians. What needs to be appreciated is that compatibility or incompatibility of two theories is determined by the compatibility criteria of a given mosaic. By *the third law*, these criteria must follow from accepted theories and other employed methods of the time. But how exactly did those criteria follow from the theories and methods of the time? When and how did they become employed? When and how were they replaced? These questions sug-

---

[58] See Lindberg (2008), pp. 226–249.

[59] See Grant (2004), pp. 220–224, 245.

[60] See Lindberg (2008), p. 250–251.

gest that there is a massive and virtually untouched layer of history that needs to be carefully studied by a theory-guided HSC.

The second aspect of *the law of compatibility* has to do with employed *methods*. When applied to methods, it basically states that two simultaneously employed methods cannot be mutually incompatible. Apparently, the notion of incompatibility of two *methods* calls for a clarification. Imagine a situation where two disciplines – say, physics and sociology – employ different requirements. Suppose that in physics a theory is considered acceptable if it provides confirmed novel predictions, while in sociology no novel predictions are required and a new theory is assessed by the number of solved conceptual and empirical problems. In this hypothetical case, can we say that the two methods are incompatible? The answer is "no", for the methods in question apply to different disciplines. Indeed, the two requirements only appear conflicting but in fact they can be simultaneously employed:

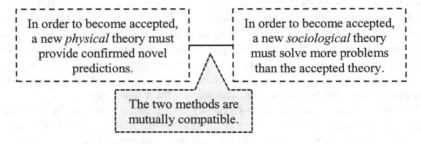

But what if the two requirements were employed in the *same* field? Would that make them incompatible? Again, the two requirements would not be mutually exclusive. In that case, we would simply have two complementary requirements: we would expect a new theory to provide some confirmed novel predictions *and* solve more problems than the accepted theory. In other words, the two requirements would be connected by a logical *AND*; we would have one method with two requirements:

> In order to become accepted, a new theory must
> (1) provide confirmed novel predictions *and*
> (2) solve more problems than the accepted theory.

The two requirements wouldn't be incompatible even if they were connected with a logical *OR*. In that case, a theory could become accepted by meeting either of the requirements. Thus, the two methods would provide two *alternative* ways for a theory to become accepted:

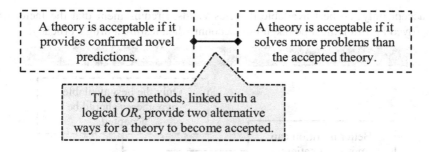

Thus, we can conclude that an apparent disparity of two requirements doesn't necessarily mean that they are incompatible, for they can be complementary (linked with an *AND*), alternative (linked with an *OR*), or pertain to different disciplines.

Two requirements are incompatible with each other only when they state *exhaustive* conditions for the acceptance of a theory. Say the first method stipulates that a theory is acceptable if and only if it provides confirmed novel predictions, while the second method requires that in order to become accepted a theory must necessarily solve more problems than the accepted theory. In this case, the two methods are incompatible and, by *the law of compatibility*, they cannot be simultaneously employed:

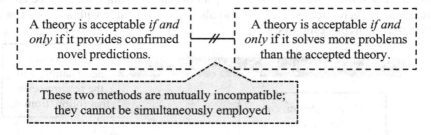

Let us now turn to the third aspect of *the law of compatibility* – the compatibility of accepted theories and employed methods. Recall that a method is a set of requirements (rules, criteria, prescriptions, etc.), whereas a theory is a set of propositions that attempt to describe/explain the world. While the former can be *explicated in* normative propositions (e.g. "a new theory ought to have confirmed novel predictions"), the latter is essentially a set of descriptive propositions. Thus, a question arises: how can a *descriptive* theory and a *prescriptive* method ever be mutually incompatible? Logically speaking, prescriptions and descriptions cannot directly conflict with each other; a method can conflict with a theory only *indirectly* by being incompatible with those methods which follow from the theory.

Consider an example. Say there is an accepted theory which says that better nutrition can improve a patient's condition. We know from the discussion in the previous section that the conjunction of this proposition with the basic requirement to accept only the best available theories yields a requirement that the factor of improved nutrition must be taken into account when testing a drug's efficacy:

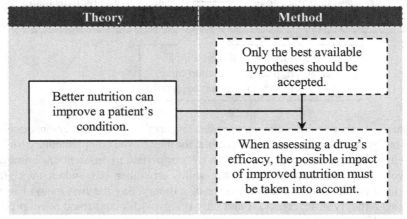

Now, envision a method which doesn't take the factor of better nutrition into account and prescribes that a drug's efficacy should be tested in a straightforward fashion by giving it only to one group of patients. This method will be incompatible with the requirement that the possible impact of improved nutrition must be taken into account. Therefore, indirectly, it will also be incompatible with a theory from which the requirement follows:

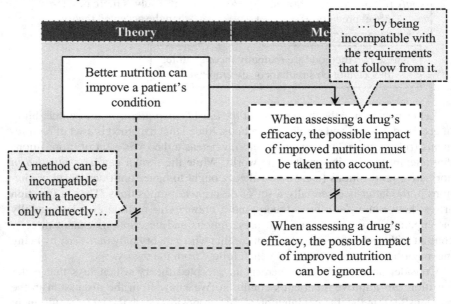

The third aspect of *the law of compatibility* covers exactly this type of incompatibility – namely, it says that a method and a theory that are mutually incompatible cannot be simultaneously in the same mosaic. I shall not elaborate on this aspect of the law for it is quite straightforward.

Chapter 5
Theorems

Having formulated the axioms of the theory, it is now time to proceed to its theorems. In this chapter, I shall take the laws as my starting point and deduce several theorems from them. Each theorem will be illustrated by relevant historical examples.

Rejection of Elements

As new elements enter into the mosaic, other elements often lose their place in it. In this section, I shall discuss several theorems concerning the mechanism of theory and method rejection. Albeit somewhat trivial, these theorems still deserve to be explicitly stated, for they will prove instrumental in our further deductions.

First, it is worth appreciating that no theory rejection (and no theory change in general) can take place in a genuinely *dogmatic* community. Namely, theory change is impossible in cases where a currently accepted theory is considered as revealing the final and absolute truth. Consider a hypothetical community that takes their currently accepted theory as the absolute and final truth. By *the third law*, this community doesn't expect any new theories on the subject. Therefore, no new theory whatsoever can meet the expectations of this hypothetical community since the community itself doesn't think that there is any need for new theories:

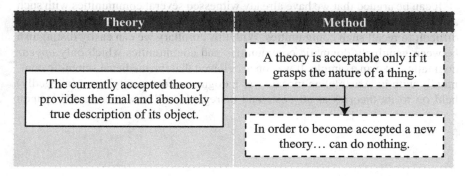

© Springer International Publishing Switzerland 2015

H. Barseghyan, *The Laws of Scientific Change*, DOI 10.1007/978-3-319-17596-6_5

In such a scenario, new theories (even if ever constructed) will have no chance to become accepted, since, by *the second law*, a new theory can become accepted only when it meets the requirements of the currently employed method, which in this case forbids any new theory acceptance. Consequently, by *the first law*, the currently accepted theory will remain accepted in the mosaic *ad infinitum*. Therefore, in genuinely dogmatic communities, there is no room for either theory acceptance or theory rejection, i.e. no theory change whatsoever. Here is the detailed deduction of the theorem:

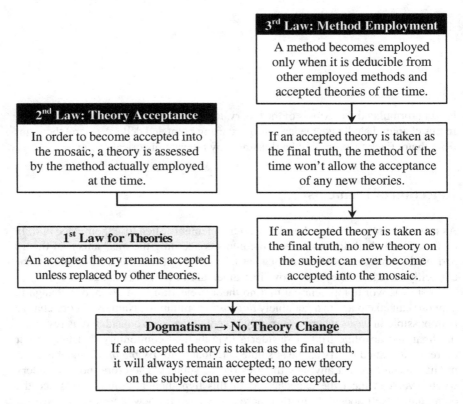

It can be argued that we have already witnessed several communities with such a dogmatic stance. I will refrain from naming names, especially given that it is not difficult to recall such communities. With this corollary we can easily distinguish between genuinely dogmatic communities and communities which only *appear* dogmatic. Take an example. It was once believed that the medieval scientific community with its Aristotelian mosaic was a dogmatic community, for it (allegedly) held on to its theories at all costs and disregarded all new theories. Yet, upon closer scrutiny it becomes obvious that the Aristotelian-medieval community was anything but dogmatic. Had the medieval community indeed taken a genuinely dogmatic stance, no scientific change would have been possible in their mosaic. But it is a historical fact that the Aristotelian-medieval mosaic was gradually changing especially in the sixteenth and seventeenth centuries; towards the end of the seventeenth century many of its key elements were replaced by new elements. Finally, by circa 1700 the Aristotelian-medieval system of theories was replaced with those Descartes and Newton. This would have been impossible had the theories of the mosaic been actually taken as revealing the final truth. Thus, the Aristotelian-medieval community was not dogmatic. For some real examples of dogmatic communities think of those communities which, having started with some dogmas, fanatically held on to those dogmas and never considered their modification possible.

This brings us to the question of *theory* rejection: under what condition does an accepted theory become rejected? By *the first law for theories*, we know that an accepted theory can become rejected only when it is replaced in the mosaic by some other theory. But *the law of compatibility* doesn't specify under what conditions this replacement takes place. For that we have to refer to *the zeroth law*, which states that at any moment of time the elements of the mosaic are mutually compatible. Suppose that a new theory meets the requirements of the time and becomes accepted into the mosaic. Question: what happens to the other theories of the mosaic? While some of the accepted theories may preserve their position in the mosaic, other theories may be rejected. The fate of an old accepted theory depends on whether it is compatible with the newly accepted theory. If it is compatible with the new accepted theory, it remains in the mosaic; the acceptance of the new theory doesn't affect that old theory in any way. This is normally the case when the new theory comes as an addition to the theories that are already in the mosaic (for instance, when the new theory happens to be the first accepted theory of its domain, i.e. when there is a new field of science that has never had any accepted theories before). Yet, if an old theory is incompatible with the new one, the old theory becomes rejected, for otherwise the mosaic would contain mutually incompatible elements, which is forbidden by *the law of compatibility*. Therefore, there is only one scenario when a theory can no longer remain in the mosaic, i.e. when other theories which are incompatible with that theory become accepted[1]:

---

[1] As with *the zeroth law*, the modified versions of both *the theory rejection theorem* and *the method rejection theorem* (discussed below) were suggested by Rory Harder during the seminar of 2013. In their initial version, both of the rejection theorems assumed that the classical logical notion of inconsistency is the universal and unchangeable criterion of compatibility. As we already know, this assumption is untenable. For details, see section "The Zeroth Law: Compatibility".

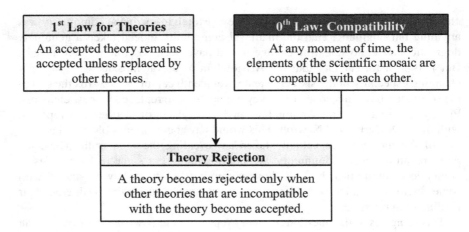

This somewhat simplistic theorem can be illustrated by a host of historical examples. Any replacement of one accepted theory by another is essentially an illustration of this theorem.

What is not trivial, however, is the specifics of the mechanism of theory rejection. Consider an oft-recurring scenario. Suppose there is an accepted theory with its axioms and theorems. Suppose that there is also a contender theory which is incompatible with the *axioms* of the accepted theory. Thus, by *the theory rejection theorem*, when the contender theory becomes accepted the axioms of the previously accepted theory become rejected. This much is clear. But what will happen with the accepted *theorems*, i.e. the theorems that followed logically from the previously accepted axioms? Will the theorems be rejected alongside the axioms or will they somehow maintain their state in the mosaic? The answer to this question is far from obvious.

Intuitively, we may be inclined to think that since they were part of the rejected theory they will be rejected too, but if we refer to *the theory rejection theorem*, it will become evident that this intuitive answer is somewhat hasty. In fact, the theorems may or may not become rejected – the fate of each theorem is decided on an individual basis by its compatibility with the propositions of the newly accepted theory. If a theorem is incompatible with the propositions of the newly accepted theory, it will be rejected. If a theorem is compatible with the propositions of the newly accepted theory, it will maintain its state in the mosaic. This will become obvious if we recall that *theory* and *proposition* are interchangeable terms, for any proposition is a folded theory, while any theory is a set of propositions or, in the extreme case, a single proposition. Thus, what the theory rejection theorem reveals to us is that a proposition is rejected only when it is replaced by other propositions that are incompatible with it. In other words, it is possible for theorems to remain accepted even when the axioms from which they followed are being rejected.

This shouldn't be surprising given that it is a simple logical truth that if the premises of a deduction are false, it doesn't necessarily entail that the conclusion is false as well. If q is a deductive consequence of p, and p turns out to be false, it doesn't entail that q is false as well. Therefore, when – for any reason – we reject the axioms of a theory, we do not always reject the theorems too.

Let us make the point more vivid by considering the historical case of *plenism*, the view that there can be no empty space (i.e. no space absolutely devoid of matter). Within the system of the Aristotelian-medieval natural philosophy, plenism was one of many theorems. Yet, when the Aristotelian natural philosophy was replaced by that of Descartes, plenism remained in the mosaic, for it was a theorem in the Cartesian system too. To appreciate this we have to consider the Aristotelian-medieval *law of violent motion*, which states that an object moves only if the applied force is greater than the resistance of the medium. In that case, according to the law, the velocity will be proportional to the force and inversely proportional to resistance. Otherwise the object won't move; its velocity will be zero[2]:

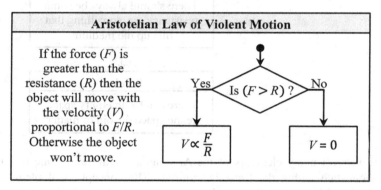

Aristotelian Law of Violent Motion

If the force (F) is greater than the resistance (R) then the object will move with the velocity (V) proportional to F/R. Otherwise the object won't move.

Is $(F > R)$?   Yes   No

$$V \propto \frac{F}{R}$$

$$V = 0$$

Taken as an axiom, this law has many interesting consequences. It follows from this law, that if there were no resistance the velocity of the object would be infinite. But this is absurd since nothing can move infinitely fast (for that would mean being at two places simultaneously). Therefore, there should always be some resistance, i.e. something that fills up the medium. Thus, we arrive at the conception of plenism[3]:

---

[2] For details, see Lindberg (2008), pp. 309–313; Cohen (1985), pp. 15–22; Dales (1973), pp. 102–108.

[3] See Lindberg (2008), pp. 54, 309–310.

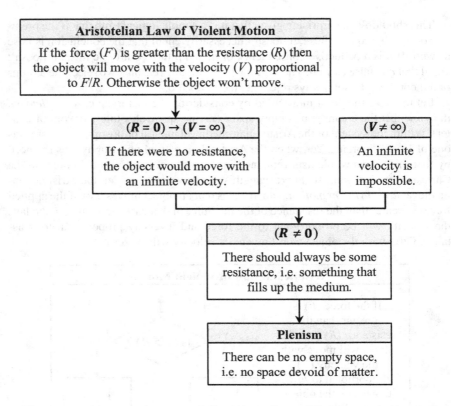

There weren't many elements of the Aristotelian-medieval mosaic that maintained their state within the Cartesian mosaic. The conception of plenism was among the few that survived through the transition. In the Cartesian system, plenism followed directly from the assumption that extension is the attribute of matter and that no attribute can exist independently from the substance in which it inheres[4]:

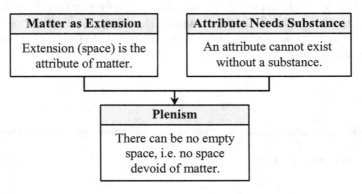

---

[4] For details, see Garber (1992), pp. 63–69, 130–136.

In short, when the axioms of a theory are replaced by another theory, some of the theorems may nevertheless manage to stay in the mosaic, provided that they are compatible with the newly accepted theory. This is essentially what *the theory rejection theorem* tells us. Thus, if someday our currently accepted general relativity gets replaced by some new theory, the theories that followed from general relativity, such as the theory of black holes, may nevertheless manage to remain in the mosaic.

The theory rejection theorem has another interesting consequence. Normally, when a theory becomes rejected, it is replaced by a new theory of the same field of inquiry. We usually expect an old physical theory to be replaced in the mosaic by a new physical theory and not, say, a new chemical or biological theory. Yet, this is not the only possible option: in order to become rejected, a theory need not be replaced by a new theory from its own field of inquiry. HSC knows several cases where an accepted theory became rejected simply because it wasn't compatible with new accepted theories of some other fields. We can formulate this point as a corollary for *the theory rejection theorem*:

Theory Rejection	Theory Rejection: Disciplines
A theory becomes rejected only when other theories that are incompatible with the theory become accepted.	A theory can become rejected not only when replaced by theories of its own discipline, but by theories of other disciplines as well.

Consider the case of *theology*'s exile from the mosaic. Note that I am not referring to theology as the study of the history of religious thought, customs, and institutions, i.e. the historical discipline which is nowadays called *religious studies*. What was exiled from the mosaic was not this historical discipline, but the one that is nowadays called *theology proper* – the study of God, his being, his attributes, and his works. Theology in this latter sense is no longer part of the mosaic. Although it is not easy to establish when exactly this exile took place, it is safe to say that, when it did, the once accepted theological propositions weren't replaced by other theological propositions. Again, it is not clear as to what exactly the accepted theological propositions were replaced with. One possible historical hypothesis is that theology was replaced in the mosaic by the thesis of *agnosticism*, an epistemological conception which is still implicit in the mosaic. Roughly, agnosticism holds that we are in no position to know about such matters as the existence of God, the attributes of God and so on. It is also possible that the exile of theology had to do with the acceptance of *evolutionary biology*.[5] In addition, it is conceivable that the exile was a gradual process: some theological propositions could have been rejected much earlier than others (apparently, the transition from proposition "God exists" to proposition "we do not know whether God exists" was the ultimate stage of this process). What seems even more likely is that form different mosaics theology was exiled for different reasons. In any

---

[5] For discussion, see Brooke (1991), pp. 275–320.

case, what must be appreciated here is that a theory can be replaced in the mosaic by theories pertaining to other fields of inquiry.[6]

The exile of *astrology* from the mosaic is yet another example. It is well known that astrology was once a respected scientific discipline and its theories were part of the mosaic. Of course, not all of the astrology was accepted; it was the so-called *natural astrology* – the theory of celestial influences on *physical* phenomena of the terrestrial region – that was part of the Aristotelian-medieval mosaic. Unlike *natural astrology*, the so-called *judicial* or *superstitious astrology*, which assumed that celestial influences affected not only the body but also the human *soul*, was unaccepted. Whereas natural astrology fit nicely into the mosaic of Aristotelian-medieval natural philosophy and theology, the judicial astrology was in conflict with the then-accepted views on God's omnipotence and human free will. In short, it was natural astrology that was taught in the European universities since the late twelfth century.[7] Although, for now, we cannot reconstruct all the details or even the approximate decade when the exile of natural astrology took place, one thing is clear: when the once-accepted theory of natural astrology became rejected, it wasn't replaced by another theory of natural astrology.[8]

Let us now turn from theory rejection to the mechanism of *method* rejection. Take a typical case. Say we have a set of accepted theories and a very simplistic method, which consists of only one requirement that can be roughly explicated as:

> In order to become accepted, a new theory must explain all known facts with more precision and accuracy than they are explained by accepted theories.

Suppose also that, as a result of changes in the accepted theories, some new method becomes employed. Question: what happens to this old method? Does it get rejected or does it still remain employed together with the new one?

The answer to this question depends on whether the two methods can be employed simultaneously. By *the zeroth law*, if the requirements of the two methods are compatible with each other, then the old method remains employed together with the new one or. Conversely, if the requirements of the two methods are incompatible, then *the zeroth law* dictates that the old method should go. Suppose that our new method has only one requirement which is:

---

[6] This is not surprising since, as I have stressed in section "Time, Fields, and Scale", disciplinary boundaries are both transient and ambiguous.

[7] See Lindberg (2008), pp. 271–277; Campion (2009), pp. 13–14, 44, 50–51.

[8] This is an interesting topic for professional historical research. *When exactly* was natural astrology exiled from the mosaic? What *theories* replaced natural astrology in the mosaic? Was it replaced by some physical theory and, if so, which one and when?

> In order to become accepted,
> a new theory must provide
> confirmed novel predictions.

Obviously, there is no conflict between this requirement and the old one – the two are complementary. Therefore, the two requirements will become simultaneously employed. The resulting method will read:

> In order to become accepted, a new theory
> must (1) explain all known facts with more
> precision and accuracy than accepted theories
> *and* (2) provide confirmed novel predictions.

But what if the new method were incompatible with the old method? Suppose our newly employed method has the following requirement:

> A theory must be accepted if it provides
> confirmed novel predictions regardless of
> whether it explains all known facts.

This new requirement voids the old requirement, for it says that explaining all known facts is not mandatory. Thus, the new method is in conflict with the old method. In this case, by *the law of compatibility*, the old method will have to go.

In addition, we can show that this is the only possible case when an employed method ceases to be employed. Indeed, by *the first law for methods*, an employed method remains employed unless it is replaced by some other method. But, as we have seen, a method can be replaced only by a method that is incompatible with it. Thus we arrive at *the method rejection theorem*:

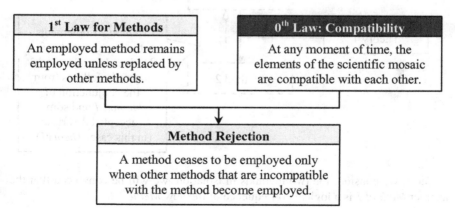

The deduction is similar to that of *the theory rejection theorem*. And just as its sibling *theory rejection theorem*, this theorem is also somewhat trivial, for any historical transition from one method to another is an illustration of this theorem.

Yet, if we now refer to *the law of method employment*, we can take a considerable step further. As we know from our discussion of *the third law*, there are two somewhat distinct scenarios of method employment. In the first scenario, a method becomes employed when it strictly follows from newly accepted theories. In the second scenario, a method becomes employed when it implements the abstract requirements of some other employed method by means of other accepted theories. It can be shown that method rejection is only possible in the first scenario; no method can be rejected in the second scenario. Namely, it can be shown that method rejection can only take place when some other method becomes employed by strictly following from a new accepted theory; the employment of a method that is not a result of the acceptance of a new theory and is merely a new implementation of some already employed method cannot possibly lead to a method rejection.

Let us start with the following case. Suppose there is a new method that implements the requirements of a more abstract method which has been in the mosaic for a while. By *the third law*, the new method becomes employed in the mosaic. Question: what happens to the abstract method implemented by the new method? The answer is that the abstract method necessarily maintains its place in the mosaic. By *the method rejection theorem*, a method gets rejected only when it is replaced by some other method which is incompatible with it. But it is obvious that our new method cannot possibly be in conflict with the old method. This is not difficult to show. To say that the new method implements the abstract requirements of the old abstract method is the same as to say that the new method follows from the conjunction of the abstract method and some accepted theories[9]:

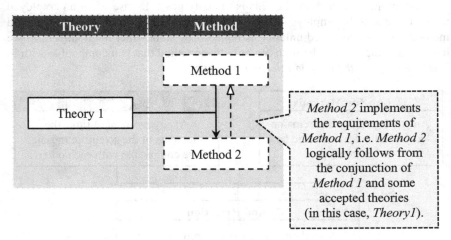

Yet, if we consider the two methods in isolation, we will be convinced that the abstract *Method 1* is a logical consequence of the new *Method 2*:

---

[9] Refer to section "The Third Law: Method Employment" for discussion.

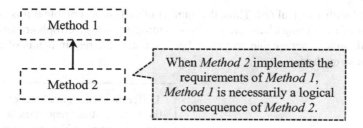

To rephrase the point, if a theory satisfies the more concrete requirements of *Method 2*, it also necessarily satisfies the more abstract requirements of *Method 1*. Recall, for instance, the abstract requirement that, when assessing a drug's efficacy, the placebo effect must be taken into account. Recall also its implementation – *the blind trial method*. It is evident that when the more concrete requirements of the blind trial method are satisfied, the more abstract requirement to take into account the possibility of the placebo effect is satisfied as well. This is because the abstract requirement is a logical consequence of the blind trial method: by testing a drug's efficacy in a blind trial, we thus take into account the possible placebo effect:

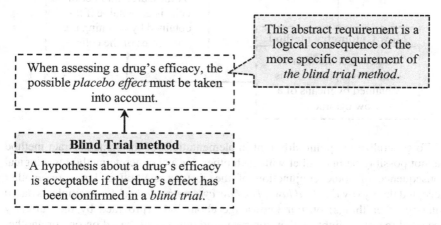

Consider another possibility. What happens when the same abstract requirement gets implemented by several concrete methods? To make the case more transparent, let us recall the history of the cell counting methods.[10] Once we understood that the unaided human eye is incapable of obtaining data about extremely minute objects (such as cells or molecules), we were led to an employment of the abstract requirement that the counted number of cells is acceptable only if it is acquired with an "aided" eye. This abstract requirement has many different implementations such as *the counting chamber method, the plating method, the flow cytometry method*, and *the spectrophotometry method*.

What is interesting from our perspective is that these different implementations are compatible with each other – they are not mutually exclusive. In fact, a researcher can pick any one of these methods, for these different concrete methods are

---

[10] The case is discussed in section "The Third Law: Method Employment".

connected with a logical *OR*. Thus, the number of cells is acceptable if it is counted by means of a counting chamber, or a flow cytometer, or a spectrophotometer. The measured value is acceptable provided that it satisfies the requirements of at least one of these methods:

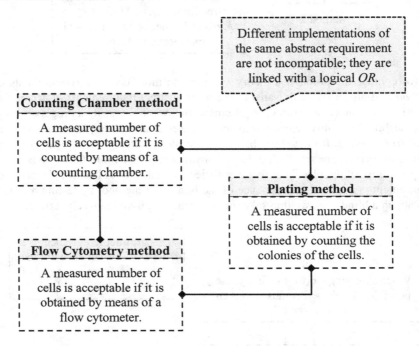

To generalize the point, different implementations of the same abstract method cannot possibly be in conflict with each other, for any concrete method is a logical consequence of some conjunction of the abstract method and one or another accepted theory (by *the third law*). *The flow cytometry method*, for example, is based (among other things) on our knowledge about light (provided by the currently accepted theory of light). Other concrete methods too are based on one or another accepted theory. In short, two implementations of the same method are not mutually exclusive and the employment of one doesn't lead to the rejection of the other. Recall that the invention and employment of *the flow cytometry method* in the 1950s–60s didn't (and couldn't) lead to the rejection of its sibling *counting chamber method*.

It is safe to take a step further and say that an employment of a new concrete method cannot possibly lead to a rejection of any other employed method. Indeed, if we take into account the fact that a new concrete method follows deductively from the conjunction of an abstract method and other accepted theories, it will become obvious that this new concrete method cannot possibly be incompatible with any other element of the mosaic. We know from *the zeroth law* that at any stage the

elements of the mosaic are compatible with each other. Therefore, no logical consequence of the mosaic can possibly be incompatible with other elements of the
mosaic. But the new method that implemented the abstract method is just one such
logical consequence. Consequently, the new employed method cannot possibly be
incompatible with other elements of the mosaic.

Since the employment of new implementations of abstract requirements cannot
lead to method rejection, there is only one place to look for a possibility of method
rejection. Recall the first scenario of method employment implicit in *the third law*.
When a new theory becomes accepted, the conjunction of that theory with the
fundamental requirement to accept only the best available theories necessarily
yields a new abstract requirement – namely that this new theory is to be taken into
account in theory assessment. Now, say we have an accepted theory which strictly
yields some abstract requirement. Question: is it possible for the requirement to
be rejected while the theory from which it follows remains in the mosaic? The
answer is "no". Since the requirement *strictly* follows from some accepted theories, its rejection is impossible without a rejection of at least some of these theories. If a theory from which the requirement follows remains in the mosaic, the
requirement will remain as well for it is a straightforward logical consequence of
the theory. This is a simple matter of logic: if p implies q, then not-q implies not-p.
In other words, the method simply cannot be rejected without a rejection of some
of the theories from which it follows. In this sense, method rejection is always
synchronous with the rejection of theories. I shall call this *the synchronism of
method rejection theorem*:

Synchronism of Method Rejection
A method becomes rejected only when some of the theories from which it follows also become rejected.

This theorem can be deduced from two premises – *the method rejection theorem*
and *the third law*. By *the method rejection theorem*, a method is rejected only when
other methods incompatible with the method become employed. Thus, we must find
out when exactly two methods can be in conflict. In order to find that out, we must
refer to *the third law* which stipulates that an employed method is a deductive consequence of accepted theories and other methods. Logic tells us that when a new
employed method is incompatible with an old method, it is also necessarily incompatible with some of the theories from which the old method follows. Therefore, an
old method can be rejected only when some of the theories from which it follows
are also rejected:

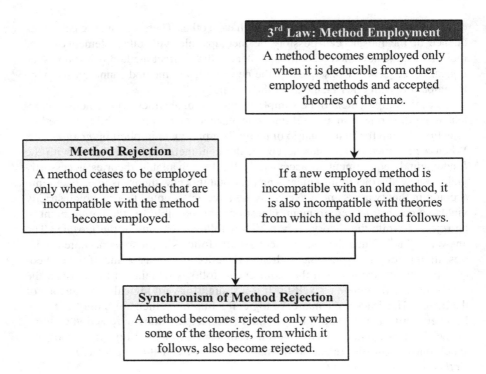

Consider some examples. When discussing the transition from *the blind trial method* to *the double-blind trial method*, it may be tempting to say that the latter came to replace the former. Although this is tolerable as a shortcut in speech, when it comes to the details of the transition it is incorrect, strictly speaking. To be sure, the blind trial method *was* replaced in the mosaic, but not by the double-blind trial method. Rather, it was replaced by the abstract requirement that when assessing a drug's efficacy one must take into account the possible experimenter's bias. The employment of the double-blind trial method was due to the fact that it specified this abstract requirement. Its employment per se had nothing to do with the rejection of the blind trial method. Let us see how this occurred.

Recall *the blind trial method* which required that a drug's efficacy is to be shown in a trial with two groups of patients, where the active group is given the real pill, while the control group is given a placebo. Implicit in *the blind trial method* was a clause that it is ok if the researchers know which group is which. This clause was based on the tacit assumption that the researchers' knowledge cannot affect the patients and, thus, cannot void the results of the trial. Although this assumption was hardly ever expressed, it is safe to say that it was taken for granted – we would allow the researchers to know which group of patients is which until we learned about the phenomenon of experimenter's bias:

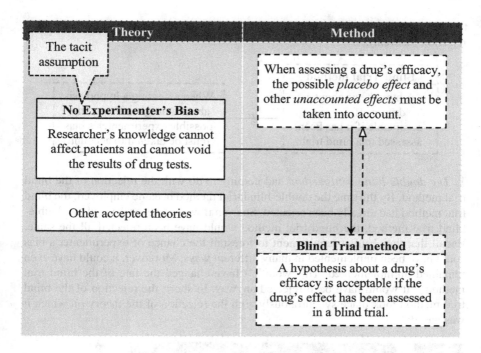

Once we learned about the possibility of experimenter's bias, the blind trial method became instantly rejected. More precisely, the acceptance of *the experimenter's bias thesis* immediately resulted in the abstract requirement that, when assessing a drug's efficacy, one must take the possibility of the experimenter's bias into account. Consequently, two elements of the mosaic became rejected: the blind trial method and the tacit assumption that the experimenters' knowledge doesn't affect the patients and cannot void the results of trials. First, *the no experimenter's bias thesis* was replaced by *the experimenter's bias thesis* (by *the theory rejection theorem*):

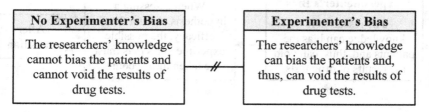

Consequently, the experimenter's bias thesis yielded the new abstract requirement to take into account the possible experimenter's bias. This requirement, in turn, replaced the blind trial method with which it was incompatible (by *the method rejection theorem*):

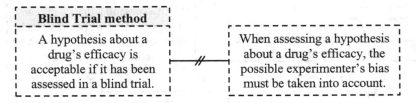

The double-blind trial method had nothing to do with the rejection of the blind trial method. By the time the double-blind trial method became employed, the blind trial method had already been rejected. So even if we had never devised the double-blind trial method, the blind trial method would have been rejected all the same. Recall that the abstract requirement to forestall the chance of experimenter's bias could have been implemented in many different ways. Moreover, it could have even remained unimplemented. This wouldn't have changed the fate of the blind trial method – it would have been rejected anyway. In short, the rejection of the blind trial method took place synchronously with the rejection of the theory on which it was based:

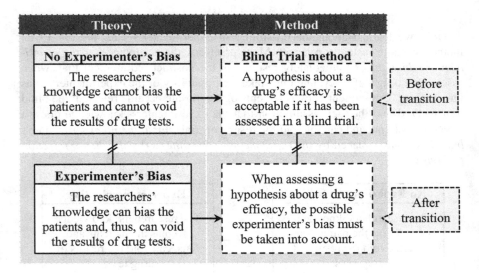

This is basically the idea of *the synchronism of method rejection theorem*.[11]

Another illustration of this theorem is provided by the rejection of the Aristotelian-medieval constraint regarding *experiments*. It was accepted in the Aristotelian-medieval mosaic that there is a strict distinction between *natural* and *artificial* things. Every natural thing, it was believed, possesses its inner source of change, its nature.

---

[11] There is an interesting theoretical question that calls for further study: is every theory rejection necessarily synchronous with some method rejection? I confess that, at the moment, I don't have a clear-cut answer to this question.

Rocks fall because it's their nature to be situated with other heavy things in the centre of the universe. Acorns grow into oak trees because that's what their nature dictates. Conversely, artificial things were believed to have an external source of change. A ship is built so that it behaves in accord with the orders of the captain. Similarly the springs and cogwheels of a clock are so constructed that they no longer behave according to their respective natures (i.e. they no longer tend to merely collect in the centre of the universe), but collectively contribute to showing the right time. In other words, when placed in artificial conditions, a thing does not behave as it is prescribed by its very nature, but as designed by the craftsman. One consequence of this distinction was the belief that the nature of a thing cannot be properly studied if it is placed in artificial conditions. It was believed that a thing cannot reveal its true nature when it is being scrutinized in an experimental set up, for any experimental set up inevitably puts the thing in artificial conditions. This is why it was accepted that experiments can reveal nothing about the natures of things. A requirement that follows from this belief is that an acceptable hypothesis that attempts to reveal the nature of a thing cannot rely on experimental data; the nature of a thing is to be discovered only by observing the thing in its natural, unaffected state. Thus, if the task is to study the nature of an animal, it is not a good idea to put it in a cage.[12]

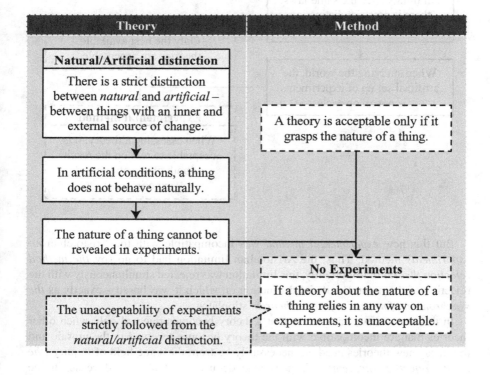

---

[12] See Lindberg (2008), pp. 49–52.

It is well known that this *no experiments* limitation was rejected during the Scientific Revolution. Importantly, the rejection was synchronous with the rejection of the natural/artificial distinction. Consider the two theories that came to replace the Aristotelian natural philosophy[13] – the Cartesian and Newtonian natural philosophies. Despite all the striking differences, the two shared many common propositions. In particular, both theories assumed that there is no strict distinction between artificial and natural, i.e. it was implicit in both theories that all material objects obey the same set of laws, regardless of whether they are found in nature or whether they are created by a craftsman. Once we accepted that there is no substantial difference between artificial and natural things, we also realized that experiments can be as good a source of knowledge about the world as observations. Consequently, we were forced to modify our method, for we could no longer neglect the experimental data:

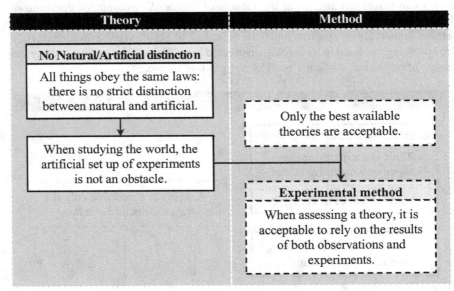

But this new *experimental method* was incompatible with the Aristotelian *no experiments* method. Thus, the Aristotelian limitation had to go (by *the method rejection theorem*). Importantly, this limitation was rejected simultaneously with the rejection of the natural/artificial distinction on which it was based – exactly as *the synchronism of method rejection theorem* stipulates.

In this section, we have learnt that a theory becomes rejected only when other theories that are incompatible with the theory become accepted in the mosaic and that these new theories need not necessarily be from the same field of inquiry (*the theory rejection theorem*). We have also deduced that a method ceases to be

---

[13] See more on this transition in section "Mosaic Split and Mosaic Merge" below.

employed only when other methods that are incompatible with the method become employed (*the method rejection theorem*). Finally, we have come to appreciate that method rejection is always synchronous with the rejection of some of the theories from which the method in question follows (*the synchronism of method rejection theorem*).

Contextual Appraisal

I shall start this section with a short historical note. Since the days of ancient philosophers and up until the mid-twentieth century, theory appraisal was generally considered as the evaluation of an *individual* theory based on the available data. We can call this *the absolute appraisal* view. It comes in two major versions – *justificationist* and *probabilist* (*neo-justificationist*).

The traditional version of *the absolute appraisal* view, championed by a majority of classical rationalists and empiricists alike, assumed that we assess a theory or even an individual proposition in order to determine whether it is true or false. Implicit to this view was the thesis of *justificationism*, i.e. the assumption that we are in a position to judge conclusively which theories are true and which are false. For *justificationists*, theory appraisal consisted of decisive proofs and equally decisive refutations.[14]

However, when it turned out that all empirical theories are equally unprovable, it became obvious that theory assessment cannot possibly consist of proofs and refutations. As a result, philosophers proposed a milder version of this view. *Probabilism* (or *neo-justificationism*, to use Lakatos's terminology) conceded that we are not in a position to decisively prove our theories, but we are in a position to measure the objective probability of an individual theory relative to the available evidence. Theory assessment, in this view, is a process whereby an individual theory is assigned a number on a probability scale. This view was proposed mainly by Cambridge philosophers Johnson, Broad, Nicod, Ramsey, Jeffreys and was later developed by Carnap, Reichenbach and other logical positivists. In one important aspect, it was akin to the traditional justificationism as it also assumed the possibility of *absolute appraisal*, i.e. the assessment of an individual theory.[15]

After *fallibilism* took off, philosophers gradually came to realise that theory appraisal cannot be absolute, for no theory can possibly be shown to be true or probable (in the objective sense).[16] Consequently, it was suggested that theory assessment is essentially a comparative procedure; what we appraise are relative merits of

---

[14] See Lakatos (1970), pp. 10–11.

[15] See Lakatos (1968) for a nice discussion of the history of *probabilism*.

[16] It is important to understand that the contemporary Bayesianism, which is essentially an heir to probabilism, has given up the task of assigning *objective* probabilities. See Howson and Urbach (2006), p. 45. Consequently, contemporary Bayesianists realise that theory assessment is a comparative procedure.

competing theories. In this *comparative appraisal* view, the goal of the assessment is the detection of the best available theory. Thus, if a theory has no extant competitors, there is very little that we can say about the theory's merits. The *comparative appraisal* view became commonly held by philosophers towards the second half of the twentieth century. Among many others, it was championed by Popper, Kuhn, Lakatos, and Laudan.[17] It is safe to say that nowadays we take this view for granted.[18]

The question that separates the comparative appraisal view from the absolute appraisal view is this:

? Does actual theory assessment concern an individual theory taken in isolation from other theories?	
Yes	**No**
The Absolute Appraisal: Actual theory assessment concerns an individual theory taken in isolation from other theories.	**The Comparative Appraisal:** Assessment of a stand-alone theory is impossible; actual theory assessment is always comparative.

Although I completely agree that all theory assessment is inevitably comparative, I believe we can go one step further. On the traditional comparativist account, all that we need for a theory assessment is two competing *theories*, some *method* of assessment, and some relevant *evidence*. Yet, if we refer to the laws of scientific change, we will see that this list is incomplete. What is missing from this list is the *scientific mosaic of the time*. What the traditional version of comparativism doesn't take into account is that, in reality, all theory assessment takes place within a specific *historical context*, i.e. within the scientific mosaic of the time.

In particular, the traditional version of comparativism holds that when two theories are compared it doesn't make any difference which of the two is currently accepted.[19] In reality, however, the starting point for every theory assessment is the current state of the mosaic. Every new theory is basically an attempt to modify the mosaic by inserting some new elements into the mosaic and, possibly, by removing some old elements from the mosaic. Therefore, what gets decided in actual theory assessment is whether a proposed *modification* is to be accepted. In other words, we judge two competing theories not in a vacuum, as the traditional version of comparativism suggests, but only in the context of a specific mosaic. It is this version of the comparativist view that is implicit in the laws of scientific change – namely, it follows from *the first* and *the second laws*. Let us demonstrate this.

---

[17] See, for instance, Popper (1934/59), pp. 281–282; Kuhn (2000), pp. 113–115; Lakatos (1970); Laudan (1977), p. 71; Laudan (1984), p. 29.

[18] See, for example, Brown (2001), p. 89; Lacey (2004), p. 12; Brock and Durlauf (1999), p. 128.

[19] This version of comparativism is implicit in Gardner (1982), p. 8; Truesdell (1960), p. 14, footnote.

By *the second law*, in actual theory assessment a contender theory is assessed by the method employed at the time. Thus, you don't go assessing the Aristotelian-medieval natural philosophy by the criteria of the twenty-first century and then wonder how on earth intelligent people can ever accept it. Such anachronisms (still abundant in popular literature) can be avoided if we realize that the theory became accepted because it managed to meet the implicit expectations of its own time.

In addition, it follows from *the first law for theories* that a theory is assessed only if it attempts to *enter* into the mosaic; once *in* the mosaic, the theory no longer needs any further appraisal (any new "confirmations", "proofs", "verifications" etc.). In this sense, the accepted theory and the contender theory are never on equal footing, for it is up to the contender theory to show that it deserves to become accepted. In order to replace the accepted theory in the mosaic, the contender theory must be declared superior by the current method; to be "as good as" the accepted theory is not sufficient. Thus, when the proponents of some alternative quantum theory argue that the currently accepted theory is no better than their own quantum theory, they take theory assessment out of its historical context. Particularly, they ignore the phenomenon of scientific inertia – they ignore that, in order to remain in the mosaic, the accepted theory doesn't need to do anything (by *the first law for theories*) and that it is their obligation to show that their contender theory is better (by *the second law*).

Such misunderstandings can be avoided if we appreciate that a theory is assessed only in the context of the mosaic of the time. I shall call this *the contextual appraisal theorem*:

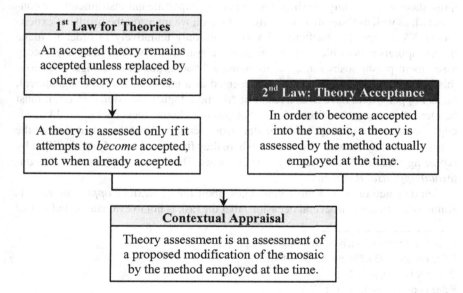

If, for whatever reason, we need to compare two competing theories disregarding the current state of the mosaic, we are free to do so, but we have to understand that in actual scientific practice such abstract comparisons play no role whatsoever. Any theory assessment always takes into account the current state of the mosaic, i.e. it always considers what theories are accepted and what methods are employed at the time of the assessment.

Some philosophers have come close to appreciating this view.[20] As early as in *The Logic of Scientific Discovery*, Popper points out that it is the modifications of a theoretical system that should be assessed, not the system itself. Surely, the central requirement of Popper's system – the requirement of falsifiability – is applicable to an individual theory, for in order to see whether a theory has any empirical content, one doesn't have to compare it with other theories. Yet, Popper realizes that falsifiability alone doesn't allow distinguishing between two competing theories when both are falsifiable and, thus, he formulates additional rules of theory appraisal which are essentially comparative. In particular, he prescribes that one must prefer that theory, which has greater empirical content, i.e. which is more falsifiable. By doing this, Popper subscribes to the traditional version of *the comparative appraisal* view. However, he inches towards *the contextual appraisal* view when he devises a rule that applies only to theory modifications: he prescribes that a theoretical system should be modified in such a fashion that the overall empirical content of the system is not diminished.[21] He comes even closer to *the contextual appraisal* view in his *Conjectures and Refutations*, where he concedes that in any experimental situation scientists "rely if only unconsciously on … a considerable amount of background knowledge".[22] In Popper's view, this background knowledge normally includes those theories according to which the respective experimental instruments are constructed, as well as those theories in light of which we interpret the results of experiments. We accept this background knowledge only tentatively in order to make theory appraisal possible.[23] It is nowadays common knowledge that any theory assessment presupposes some "unproblematic" background knowledge. However, this background knowledge is often presented as a matter of choice or agreement, or, as Popper would have it, as a result of "methodological decisions".[24] Yet, it must be clear that new generations of scientists do not choose their background knowledge, for they are in no position to start from scratch. What they deal with is the existing scientific mosaic: they take it where they find it and try only to modify it by replacing some of its elements by new elements. This idea is expressed in *the contextual appraisal theorem*.

Another author who almost converges upon *the contextual appraisal* view is Kuhn, who stresses on several occasions that the task is not to evaluate an individual

---

[20] See Musgrave (1974), p. 8; Hudson (2007), p. 17.

[21] See Popper (1934/59), pp. 32–33, 61–63.

[22] Popper (1963), p. 322.

[23] See Popper (1963), pp. 151, 322–330.

[24] Giere sounds along these lines in his (1984), pp. 20, 23. For discussion, see Lakatos (1970), pp. 23–31.

belief (for we are incapable of doing so), but to evaluate the *change* of belief.[25] However, Kuhn doesn't quite subscribe to *the contextual appraisal* view for, in Kuhn's conception, accepted theories and employed methods (*the mosaic* in my vocabulary) play a role in theory assessment only during the so-called periods of normal science. As for the role of the mosaic in theory assessment during the so-called scientific revolutions, Kuhn's view is extremely ambiguous.[26]

In Lakatos's conception, theory assessment is always an evaluation of modifications; his famous three rules apply to *modifications* in a research programme. This brings Lakatos very close to the contextual appraisal view. However, he isn't quite there either, since he doesn't see any significant difference between accepted and unaccepted research programmes, for according to Lakatos scientists evaluate not only modifications in the mosaic of accepted theories and employed methods, but modifications in any research programme. In his view, two research programmes are always on equal footing regardless of which of the two is currently accepted. The assessment of research programmes is merely a means to know which one is currently more progressive, i.e. a means of keeping score.[27]

Thus, it would be fair to say that although many theoreticians have come close to accepting *the contextual appraisal view*, the view itself is yet to be fully appreciated.

To illustrate the key point of *the contextual appraisal theorem*, I shall refer to the case of Galileo's heliocentrism. Until recently, this case was often presented as an illustration of a scientist-genius fighting against the ignorance, dogmatism, and irrationality of his contemporaries, of a hero struggling to overthrow an obsolete intellectual tradition. Galileo's position was traditionally portrayed as though it was clearly superior to that of the Church authorities. As a result, the dismissal of Galileo's views by the Church authorities was considered extremely unjust and Galileo himself was depicted as a victim of religious persecution. This view was nicely expressed by Albert Einstein, who characterized Galileo as a man who possessed "the passionate will, the intelligence, and the courage to stand up as the representative of rational thinking against the host of those who, relying on the ignorance of the people and the indolence of teachers in priest's and scholar's garb, maintain and defend their positions of authority."[28] This traditional interpretation

---

[25] See Kuhn (2000), pp. 112–115.

[26] Although in his (1977), pp. 320–339, Kuhn attempted to introduce five transhistorical *values* which presumably guide theory assessment across paradigms, it is clear that the attempt was a flop, for he ended up making these values paradigm-dependent. For discussion, see Laudan (1984), pp. 14–20, 30–32.

[27] See Lakatos (1970), pp. 32–52.

The closest thing to *the contextual appraisal* view that I have come across in the literature is the view expressed by Brown in his (2001), pp. 108, 131, 141 and, especially, 158–159.

[28] Einstein (1953), p. vii.

Unfortunately the vices of the traditional interpretation are still being repeated in popular accounts. Take for instance Mark Steel's BBC lectures. Albeit ingeniously hilarious, they propagate the same old errors by presenting the case as if it were "the brilliance of a few" versus "the dogmatism of churchmen".

was quite straightforward: we know that geocentrism is wrong; therefore, the Church authorities were wrong in maintaining geocentrism.

Albeit heroic, this traditional interpretation was obviously flawed since it didn't take into account the *scientific mosaic* of the time, it carried the comparison of Galileo's heliocentrism with the Aristotelian-Ptolemaic geocentrism in a vacuum. The traditional account failed to appreciate both that theory assessment is an assessment of a proposed modification and that a theory is assessed by the method employed at the time. Once we focus our attention on the state of the scientific mosaic of the time, once we realise what theories were accepted and what methods were employed in the early-seventeenth century, it becomes obvious that the scientific community of the time simply couldn't have acted differently.

Let us start from the state of the mosaic circa the 1610s and see what it included. It consisted of many interconnected elements such as the Aristotelian-medieval theories of form and matter, of four kinds of motion, of four causes, of four terrestrial elements and so on. *Geocentrism*, the view that the Earth is in the centre of the universe, was a deductive consequence of the Aristotelian *law of natural motion* and the theory of *elements*. According to the Aristotelian law of natural motion, all terrestrial elements in their natural state tend towards their natural position and remain there once that position is reached. The natural position of heavy elements is the centre of the universe, whereas the natural position of light elements is the periphery of the sublunar (terrestrial) region. *Earth* and *water*, as heavy elements, tend towards the centre of the universe. And since *earth* is heavier than *water*, earth ends up in the very centre, while water constitutes the next layer. Thus, according to the then-accepted view, the Earth that consists of elements *earth* and *water* should necessarily be located in the centre of the universe[29]:

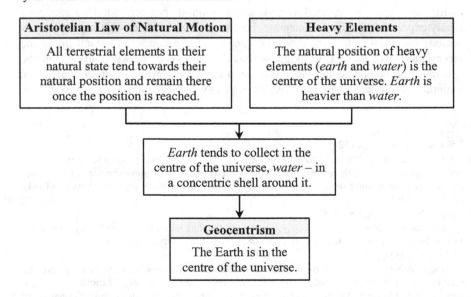

Aristotelian Law of Natural Motion	Heavy Elements
All terrestrial elements in their natural state tend towards their natural position and remain there once the position is reached.	The natural position of heavy elements (*earth* and *water*) is the centre of the universe. *Earth* is heavier than *water*.

Earth tends to collect in the centre of the universe, *water* – in a concentric shell around it.

Geocentrism
The Earth is in the centre of the universe.

---

[29] See Kuhn (1957), pp. 55–60, 79–85; Lindberg (2008), p. 55.

So it was impossible to simply cut geocentrism out of the mosaic and replace it with heliocentrism – the whole Aristotelian theory of elements would have to be rejected as well. But the difficulty was that the theory of elements itself was tightly connected with many other parts of the mosaic. Among other things, it was tightly linked to the then-accepted views on the possibility of transformation of elements. It also provided a foundation for the *medical* theory of the time: there was an obvious connection between the four elements and the four basic constituents of the body, four *humors*.[30] Consequently, the existence of four elements was assumed by the then-accepted theory of *human temper*, where the difference between temperaments was explained in terms of the predominance of one of the humors (and, thus, one of the elements). The idea of four elements also played an important role in the accepted *astrological* theory of the time. In short, in order to make the rejection of geocentrism possible, a whole array of other elements of the Aristotelian-medieval mosaic would have to be rejected as well.

No doubt, Galileo managed to devise ingenious arguments against geocentrism. But as we know from *the first law for theories* and *the theory rejection theorem*, only the acceptance of an alternative set of theories could defeat the theories of the Aristotelian-medieval mosaic. Unfortunately for Galileo, at the time (in the first third of the seventeenth century), there was no acceptable contender theory comparable in scope with the theories of the Aristotelian-medieval mosaic. All his efforts notwithstanding, Galileo didn't have an acceptable replacement for all the elements of the mosaic that had to be rejected together with geocentrism. The situation changed only towards the middle of the century, when Descartes constructed his system of natural philosophy where the Earth was one of the planets of the solar vortex. Prior to that, heliocentrism had no chance whatsoever of becoming accepted. The traditional interpretation of this historical episode failed to appreciate this important point and, instead, preferred to blame the dogmatism of the clergy.

Another fault of the traditional interpretation was that the whole episode was assessed not by the implicit requirements of the time, but by the requirements of the hypothetico-deductive method, which became actually employed a whole century after the episode took place. Namely, Galileo was said to have shown the superiority of the Copernican heliocentrism by confirming some of its *novel predictions*. It is well known that the Copernican theory provided several novel predictions, such as stellar parallax and the phases of Venus and Mercury. According to the traditional view, Galileo's telescopic observations confirmed some of these predictions. Indeed, in 1610, Galileo observed a full set of Venus's phases. The traditional account of the episode considered this a clear-cut indication of the superiority of the Copernican hypothesis – a superiority the clergy failed to see due to their obstinacy.

Yet, a more careful study of the episode reveals that the requirements of hypothetico-deductivism had little in common with the actual expectations of the community of the time. Although the task of reconstructing the late Aristotelian-

---

[30] See Lindberg (2008), pp. 115–117, 339–342.

medieval method of natural philosophy is quite challenging and may take a considerable amount of labour, one thing is clear: the requirement of confirmed novel predictions was not among implicit expectations of the community of the time. Back then, theories simply didn't get assessed by their confirmed novel predictions.[31] This becomes obvious once we appreciate *the contextual appraisal theorem* and start paying attention to the mosaic of the time.

To provide another illustration for *the contextual appraisal theorem*, I shall refer to the famous Eucharist episode which took place in the second half of the seventeenth century. The history of the transition from the Aristotelian-medieval natural philosophy to the Cartesian and Newtonian natural philosophies around the year 1700 is arguably one of the most difficult cases to account for by any theory of scientific change. One difficulty that has to be dealt with is that the transition was not uniform: unlike a great majority of other transitions, the accepted theory (the Aristotelian-medieval natural philosophy) was replaced by not one, but two new theories. While in Cambridge or Paris it was replaced by the Cartesian natural philosophy, in Oxford or Edinburgh it was replaced by the Newtonian natural philosophy. Nor did the transition take place simultaneously in different regions. While in Cambridge the Cartesian natural philosophy became accepted circa 1680, in Paris it became accepted only circa 1700. The situation becomes even more complex when we recall that the transition also involved changes in employed methods. This complex transition raises several important questions. How is it possible for two competing theories to be simultaneously accepted by different scientific communities? More generally, how can one scientific mosaic split into two or more scientific mosaics? And if the mosaics do in fact split, then how do they merge again? I shall address the questions of splitting and merging of mosaics in due course.[32] Here, I would like to focus on a more subtle issue: on the role of the Eucharist episode in delaying the acceptance of the Cartesian natural philosophy in Paris.

This episode has been often portrayed as a clear illustration of how religion affects science. In particular, the episode has been presented as though the acceptance of Cartesianism in Paris was delayed due to the role played by the Catholic Church. It is a historical fact that Descartes's natural philosophy was harshly criticized by the Church. In 1663, his works were even placed on the *Index of Prohibited Books* and in 1671 his conception was officially banned from schools. Thus, at first sight, it may appear as though the acceptance of the Cartesian science in Paris was indeed hindered by religion. Yet, upon closer scrutiny, it becomes obvious that this interpretation is too superficial.

When Descartes constructed his natural philosophy, it soon turned out that it had a very troubling consequence: it wasn't readily reconcilable with *the doctrine of transubstantiation* accepted by the Aristotelian-Catholic scientific community of Paris. The idea of transubstantiation was proposed by Thomas Aquinas in his

---

[31] For my explication of the Aristotelian-medieval method, see pp. 139 ff.

[32] See section "Mosaic Split and Mosaic Merge" below.

Summa Theologiae as an explanation of one of the Christian dogmas – namely, that of the *Real Presence* which states that, in the Eucharist, Christ is really present under the appearances of the bread and wine (i.e. literally, rather than metaphorically or symbolically)[33]:

Real Presence
In the Eucharist, Christ is really present in the bread and wine.

In his explanation of *Real Presence*, Aquinas employed Aristotelian concepts of *substance* and *accident*. In particular, he stated that in the Eucharist the consecration of bread and wine effects the change of the whole substance of the bread into the substance of Christ's body and of the whole substance of the wine into the substance of his blood. Thus, what happens in the Eucharist is *transubstantiation* – a transition from one substance to another. As for the accidents of the bread and wine such as their taste, color, smell etc., Aquinas held that they remain intact, for transubstantiation doesn't affect them. The doctrine of transubstantiation soon became the accepted Catholic explanation of the Real Presence. Its acceptance was reaffirmed in 1551 by the Council of Trent[34]:

Transubstantiation
In the Eucharist, the consecration of the bread and wine effects the change of the whole substance of the bread and wine into the substance of Christ's body and blood; only the accidents (smell, taste, color etc.) remain intact.

The problem was that Descartes's theory of matter didn't provide any mechanism similar to that stated in *the doctrine of transubstantiation*. To be more precise, it followed from Descartes's original theory that transubstantiation was impossible. Recall that, according to Descartes, the only principal attribute of matter is extension: to be a material object amounts to occupying some space. It follows from this basic axiom that accidents such as smell, color, or taste are effects produced upon our senses by the configuration and motion of material particles. In other words, we simply cannot perceive the accidents of bread and wine unless there is bread and wine in front of us. What makes bread what it is, what constitutes its substance (to use Aristotle's terms) is a specific combination of material particles; and the same goes for wine. Thus, when the substance of bread changes into the substance of Christ's body, in the Cartesian theory, it means that some combination of particles

---

[33] See Grant (2004, p. 216; Mathews (2008), p. 67; Bourg (2001), p. 122.
[34] See Bourg (2001), p. 121; Nadler (1988), p. 231; Mathews (2008), p. 67. Recently, it was reaffirmed by John Paul II. See John Paul II (2003), paragraph 15.

which constitutes the bread changes into another combination of particles which constitutes Christ's body. The key point here is that, in Descartes's theory, it is impossible for Christ's body to have the appearance of bread, since the appearance is merely an effect produced by that specific combination of particles upon our senses; Christ's body and blood simply cannot produce the accidents of bread and wine. Obviously, on this point, Descartes's theory was in conflict with *the doctrine of transubstantiation*[35]:

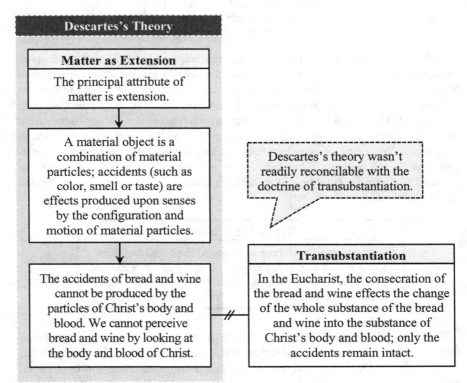

This conflict became the focal point of criticism of Descartes's theory. To a twenty-first-century reader used to a clear-cut distinction between science and religion this may seem a purely religious matter. Yet, in the second half of the seventeenth century, this was precisely a scientific concern. The crucial point is that back then theology wasn't separate from other scientific disciplines: the scientific mosaic of the time included many theological propositions such as "God exists", "God is omnipotent", or "God created the world". These propositions where part of the mosaic just as any other accepted proposition. If we could visit seventeenth-century Paris, we would see that *the dogma of Real Presence* and *the doctrine of transubstantiation* weren't something foreign to the scientific mosaic

---

[35] See Pagden (1988), p. 130; Nadler (1988), p. 230; Mathews (2008), p. 69.

of the time – they were accepted parts of it alongside such propositions as "the Earth is spherical", "there are four terrestrial elements", "there are four bodily fluids" and so on. Thus, Descartes's theory was in conflict not with some "irrelevant religious views" but with a key element of the scientific mosaic of the time, *the doctrine of transubstantiation*:

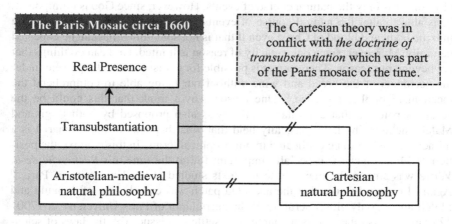

More precisely, the problem was that back then no theory was allowed to be in conflict with the accepted theological propositions. This latter requirement was part of the method of the time. The requirement strictly followed from the then-accepted belief that theological propositions are infallible:

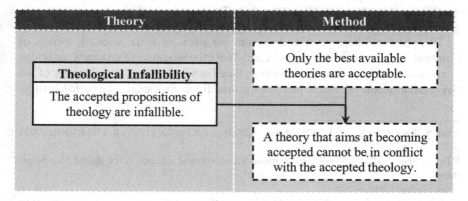

Yet, eventually, the Cartesian natural philosophy did become accepted in Paris. If the laws of scientific change are correct, it could become accepted only with a special patch that would reconcile it with the doctrine of transubstantiation. It is not clear as to what exactly this patch was. To be sure, there is vast literature on different Cartesian solutions of the problem: the solutions proposed by Descartes,

Desgabets, and Arnauld are all well known.[36] However, I have failed to find a single historical narrative revealing which of these patches became accepted in the mosaic alongside the Cartesian natural philosophy circa 1700.[37] Based on the available data, I can only hypothesize that the accepted patch was the one proposed by Arnauld in 1671. According to Arnauld's solution, the Cartesian natural philosophy concerns only the natural course of events. However, since God is omnipotent, he is able to alter the natural course of events. Thus, he can turn bread and wine into the body and blood of Christ even if that is not something that can be expected naturally. Moreover, since our capacity of reason is limited, God can do things that are beyond our reason. Therefore, it is possible for Christ to be really present under the accidents of the bread and wine without our being able to comprehend the mechanism of that presence.[38] One reason why I think that this could be the accepted patch is that a similar solution was also proposed by both Régis and Malebranche.[39] The latter basically held that what happens in the Eucharist is a miracle and is not to be explicated in philosophical terms. In this context, the position of Malebranche is especially important for, at the time, his *Recherche de la Vérité* was among the main Cartesian texts studied at the University of Paris.[40] Again, I cannot be sure that the accepted patch was exactly that of Arnauld and Malebranche; only closer scrutiny of the curriculum of Paris University in 1700–1740 as well as other relevant sources can settle this issue. Yet, the laws of scientific change tell us that there should be one patch or another – the Cartesian natural philosophy couldn't have been accepted without one.

In short, initially the Cartesian theory didn't satisfy the requirements implicit in the mosaic of the time, namely it was in conflict with one of those propositions which were not supposed to be denied. Thus, the acceptance of Descartes's theory was hindered not because "dogmatic clergy" didn't like it on some mysterious religious grounds, but because initially it didn't satisfy the requirements of the time.

This point will become clear if we turn our attention to the scientific mosaic of Cambridge of the same time period. Circa 1660, the mosaics of Paris and Cambridge were similar in many respects. For one, they both included all the elements of the Aristotelian-medieval natural philosophy. In addition, they shared the basic Christian

---

[36] See Nadler (1988), pp. 233–242; Schmaltz (2005), p. 86; Pagden (1988), p. 130; Easton (2005), p. 29.

[37] This is another illustration of what happens when historical research is not guided by a proper theory.

[38] See Nadler (1988), p. 241.

It is worth pointing out that Arnauld didn't exactly invent this patch; he merely exploited a well-known medieval strategy. In fact, some 400 hundred years before Arnauld, a very similar strategy was used to reconcile the Aristotelian natural philosophy with Christian dogmas such as "God's omnipotent" or "God created the world". See Lindberg (2008), pp. 240–243, 251–253; Grant (2004), pp. 216–217.

[39] See Schmaltz (2005), p. 86; Schmaltz (1996), p. 267, footnote 28.

[40] See Vartanian (1953), p. 42.

dogmas, such as *the dogma of Real Presence*. Yet, they were different in one important respect: whereas the mosaic of Paris included the propositions of Catholic theology, the mosaic of Cambridge included the propositions of Anglican theology. Namely, the Cambridge mosaic didn't include *the doctrine of transubstantiation*. In that mosaic, the Cartesian theory was only incompatible with the Aristotelian-medieval natural philosophy which it aimed to replace:

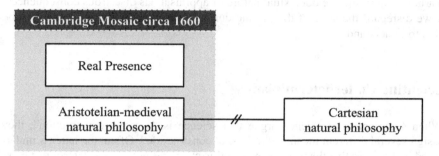

This difference proved crucial. Whereas reconciling the Cartesian natural philosophy with *the doctrine of transubstantiation* was a challenging task, reconciling it with *the dogma of Real Presence* wasn't difficult. One such reconciliation was suggested by Descartes himself and was developed by Desgabets. The idea was that the bread becomes the body of Christ by virtue of being united with the soul of Christ, while the material particles of the bread remain intact. For the Catholic, this solution was unacceptable, for it denied *the doctrine of transubstantiation* and, therefore, was a heresy. Yet, for the Anglican, this solution could be acceptable, since *the doctrine of transubstantiation* wasn't part of the Anglican mosaic. Thus, whereas the Catholic was faced with a seemingly insurmountable problem of reconciling the Cartesian natural philosophy with the doctrine of transubstantiation, the Anglican didn't have that problem. This explains why the whole Eucharist case was almost exclusively a Catholic affair.

This episode illustrates the main point of *the contextual appraisal theorem*: a theory is assessed only in the context of a specific mosaic and the outcome of the assessment depends on the state of the mosaic of the time. No doubt, some sociologically oriented authors would readily qualify this as an instance of reception that depended on cultural factors (e.g. local attitudes towards religion or politics).[41] I disagree with that interpretation. What sociologists would improperly qualify as "religious influence" was merely a set of accepted theological *propositions* of the scientific mosaic and employed *requirements* which followed from them. I agree with sociologists that the process of scientific change must be seen in its historical context, but most of what they include in their "historical context" is in fact part of the mosaic of the time – a set of accepted propositions and employed methods. That

---

[41] See, for instance, Fara (2003), pp. 488, 491–492.

is why it is advisable to stop talking about vague and elusive "sociocultural contexts" and rather focus on detailed reconstructions of the respective mosaics.[42]

In this section, we have discussed *the contextual appraisal theorem*. According to the theorem, assessing a theory amounts to assessing a proposed modification of the method that was actually employed at the time of the assessment. This theorem helps to avoid anachronisms still present in some historical narratives. We have learnt that ignoring the contextual nature of appraisal has disastrous consequences: if we disregard the state of the mosaic during a particular transition, we won't be able to understand it.

Scientific Underdeterminism

When sociologists of science argue that science is "a social construction", they assign several different meanings to "social construction". Often, the latter is understood in the sense that the process of scientific change is not strictly deterministic, or, to use my terminology, that transitions from one state of the mosaic to the next are not always inevitable.[43] I shall call this the thesis of *scientific underdeterminism*:

Scientific Underdeterminism
Transitions from one state of the mosaic to another are not necessarily deterministic. Scientific change is not a strictly deterministic process.

This thesis has had adherents not only among sociologists but also philosophers. Kuhn, for instance, pointed out in *The Essential Tension* that the criteria employed in theory assessment do not always guarantee a conclusive outcome.[44] The later Laudan is of the same opinion. It is one of the key tenets of his *reticulated model* that neither accepted theories nor employed methods are strictly determined, since theories and methods underdetermine each other. Thus, according to the reticulated model, science could have evolved differently – if we had two different scientific communities, the successive states of their respective

---

[42] See section "Sociocultural Factors" below for discussion.

[43] See, for instance Biagioli (1996), p. 201.

[44] See Kuhn (1977), pp. 290–291, 322–328. I should mention, however, that none of the cases discussed by Kuhn is a genuine instance of inconclusiveness. What Kuhn ignored in his examples (such as Ptolemy vs. Copernicus, or phlogiston vs. oxygen) is that, in actual practice, theories are assessed by the method *of the time* and not by Kuhn's own five criteria. Once we assess Kuhn's cases by the methods employed at their respective times, we come to realize that the examples were chosen erroneously.

mosaics could be very different.[45] Brown shares this view: "it is extremely unlikely that two unconnected intellectual communities would have identical histories."[46]

The thesis of *scientific underdeterminism*, as defined here, is opposed to a more traditional view that scientific change is a strictly deterministic process.[47] This view, which can be called *scientific determinism*, was tacitly assumed by many generations of philosophers. Laudan traces it back to Leibniz,[48] although it would be safe to say that perhaps Plato and Aristotle would also readily embrace it. In any case, the thesis of scientific determinism was shared by many philosophers, including logical positivists, as well as Popper and Lakatos. It has its proponents even nowadays. Weinberg, for instance, holds that two independent scientific communities must necessarily end up discovering the same laws of nature.[49]

Let us formulate the question that separates the two parties:

? Is scientific change a strictly deterministic process? Will the histories of two unconnected scientific mosaics be necessarily similar?	
Yes	**No**
Scientific Determinism: Transitions in the mosaic are strictly deterministic. Two unconnected scientific communities will necessarily pass through similar stages.	**Scientific Underdeterminism:** Transitions in the mosaic are not strictly deterministic. Two unconnected scientific communities won't necessarily pass through similar stages.

Presently, dismissing *scientific determinism* may seem a somewhat simplistic task. Indeed, it is accepted in the contemporary psychology that the process of theory construction takes a fair amount of creativity and imagination. Contrary to the hopes of classical empiricists, theories are no longer believed to be "deducible from phenomena", they are not "discovered" in any way. Thus, there was nothing inevitable in our acceptance of, say, the Galenic medicine, the Cartesian natural philosophy, or the phlogiston theory, for the very invention of these theories wasn't inevitable. In short, a theory may or may not be constructed – nothing necessitates that it inevitably will be. Therefore, it can be concluded that scientific determinism is unacceptable. This conclusion can be secured even without any TSC. Yet, the TSC allows us to approach the issue from a somewhat different angle. It is easy to notice that the thesis of scientific underdeterminism is one of the theorems of the

---

[45] See Laudan (1984), pp. 26–39, 43–45, 62.

[46] Brown (2001), p. 20.

[47] *Underdeterminism* in this wider sense should not be confused with a more specific notion of *underdetermination of theory by evidence*. For a clarification of the latter, see Laudan and Leplin (1991) and Stanford (2013).

[48] See Laudan (1984), pp. 5, 11, 25, 33.

[49] See Weinberg (2003), p. 150.

theory. The laws of scientific change make it obvious that neither theory change nor method change are strictly deterministic processes. I shall consider them in turn starting with a more simple case – *the underdeterminism of method change*.

Recall *the third law*, which allows for two distinct scenarios of method employment. In the first scenario, a method becomes employed when it strictly follows from the accepted theories and other employed methods. In the second scenario, a method becomes employed when it implements the abstract requirements of some other employed method. It is readily seen that the process of method employment is strictly deterministic only in the first scenario of method employment. Indeed, when a method is a straightforward logical consequence of accepted theories and other methods, its employment is inevitable. Yet, in the second scenario, there are many possible courses of events, since the same abstract requirements can be implemented in many different ways, given sufficient ingenuity. As I have explained in section "The Third Law: Method Employment", any set of abstract requirements can, in principle, receive an infinite number of implementations. This gives room for human creativity and genuine innovation. Thus, the methods that we end up employing are by no means the only possible implementations of the abstract requirements that follow from our accepted theories; it is quite conceivable that the same requirement could have been implemented by several very different methods and, thus, we could have ended up employing methods very different from what we actually employ. Therefore, it should be concluded that the process of transitions from one employed method to another is not strictly deterministic. This is an immediate corollary of the third law:

3rd Law: Method	Underdetermined Method Change
A method becomes employed only when it is deducible from other employed methods and accepted theories of the time.	The process of method change is not necessarily deterministic: employed methods are by no means the only possible implementations of abstract requirements.

Recall the case of the cell counting methods. It follows from our accepted theories that, when counting the number of living cells, the resulting value is acceptable only if it is obtained with an "aided" eye. This abstract requirement has been implemented by many different methods such as *the plating method*, *the counting chamber method*, *the flow cytometry method* etc. Similarly, *the double-blind trial method* is not the only logically possible implementation of the abstract requirements that it implements, just as *the Aristotelian-medieval method* wasn't the only implementation of the abstract requirements of the time. Even our *hypothetico-deductive method* is not the only possible implementation of the abstract requirements that follow from our accepted theories.[50]

The case of *theory* change is somewhat more complex. Consider the usual situation where there is a mosaic with some accepted theories and employed methods

---

[50] See section "The Third Law: Method Employment" for discussion.

and there is also a contender theory that aims at becoming part of the mosaic. *The second law* tells us that, in order to become part of the mosaic, the theory is assessed by the respective employed method. But what does it mean for a theory to be assessed by the employed method? We know that employed methods are merely our actual expectations regarding new theories (by the definition of *employed method*). Thus, when we face a new theory, we basically assess the theory by confronting it with our expectations. This process of theory assessment may have three possible outcomes – two *conclusive* and one *inconclusive*. The two conclusive outcomes are obvious – "accept" and "do not accept". While in the former case the method tells us that the theory must necessarily be accepted, in the latter case it prescribes that the theory must remain unaccepted. However, from a logical point of view, not all possible outcomes of theory assessment are necessarily conclusive. It is conceivable that theory assessment may have an inconclusive outcome. In such an instance, neither the acceptance of the theory nor its unacceptance is obligatory.

Consider an example. Suppose there is a new theory which we confront with our expectations, and our gut feeling tells us "alright, it looks like a decent theory, but it could have been better…". In other words our employed method prescribes something like "a theory can be accepted". But it is obvious that this is not the kind of outcome that can be called conclusive. In this case, the outcome of theory assessment is inconclusive, since "can" merely indicates a *possibility* and not a *necessity*. This inconclusiveness may be due to many different factors, such as the nature of the theory under scrutiny, the available evidence, the state of the mosaic, the peculiarities of the method itself (or any combination of these factors). In any case, what is important in this context is that, from a logical perspective, a conclusive outcome is not something guaranteed. We must acknowledge that there are not two but three logically possible outcomes of theory assessment:

Outcome: Accept ≡
An outcome of theory assessment which prescribes that the theory must be accepted.

Outcome: Inconclusive ≡
An outcome of theory assessment which allows for the theory to be accepted but doesn't dictate so.

Outcome: Not Accept ≡
An outcome of theory assessment which prescribes that the theory must not be accepted.

Let us first consider the case with one contender theory that undergoes assessment. When an outcome of theory assessment is a conclusive "must be accepted", the theory under scrutiny becomes accepted (by *the second law*). When an outcome is a conclusive "must remain unaccepted", the theory remains unaccepted (again by *the second law*). The result is unambiguous in both case and the future state of the mosaic is strictly determined. Yet, when an outcome of theory assessment is inconclusive, the future of the mosaic is no longer strictly determined. Indeed, when an outcome is "can be accepted", the community can choose to accept the theory or it can prefer to leave the theory unaccepted.[51]

Consider a hypothetical mosaic where the employed method prescribes the following:

> In order to become accepted, a new theory must solve
> more important problems than the accepted theory.

Despite the early Laudan's enthusiasm, it is clear that this requirement is vague: it is not always easy to tell which theory solves more problems or which problems are more important. Moreover, it is often not transparent as to what constitutes a genuine problem and what does not. Now, suppose there is a contender theory which is assessed by this requirement. It is quite conceivable that the outcome of the assessment might be inconclusive. It may turn out, for instance, that some problems are better solved by the new theory, while others are better solved by the old accepted theory. It may also turn out that the new theory solves some problems which do not even count as problems from the perspective of the accepted theory and vice versa. Many other similar scenarios are possible. Therefore, it may easily turn out that we are in no position to tell whether our implicit requirements were satisfied; the outcome of our assessment may be inconclusive.

Now, whether there have been any historical cases of inconclusive theory assessment is a factual issue that must be addressed by HSC. Although I think we have enough evidence to claim that at least some historical theory assessments have been inconclusive, we do not need to settle that issue here. The crucial point is not whether *so far* there have been any actual cases of inconclusive theory assessment, but the very *possibility* of inconclusiveness. Since theories are accepted only after being assessed by the current method (by *the second law*) and since theory assessment may have an inconclusive outcome (by definition), we can conclude that the process of theory change is not necessarily deterministic:

2nd Law: Theory Acceptance	**Theory Assessment Outcomes ≡**
In order to become accepted into the mosaic, a theory is assessed by the method actually employed at the time.	The outcome of theory assessment is not necessarily conclusive; an inconclusive outcome ("*can* be accepted") is also conceivable.

Underdetermined Theory Change
The process of theory change is not necessarily deterministic: there may be cases when both a theory's acceptance and its unacceptance are equally possible.

---

[51] There is also the third option, which I shall discuss in section "Mosaic Split and Mosaic Merge".

If we recall that the process of method change is also not always deterministic, we will arrive at the thesis of *scientific underdeterminism*:

Underdetermined Method Change	**Underdetermined Theory Change**
The process of method change is not necessarily deterministic: employed methods are by no means the only possible implementations of abstract constraints.	The process of theory change is not necessarily deterministic: there may be cases when both a theory's acceptance and its unacceptance are equally possible.

Scientific Underdeterminism
Transitions from one state of the mosaic to another are not necessarily deterministic. Scientific change is not a strictly deterministic process.

But let us return to theory change for a moment. Are there any actual examples of inconclusive theory assessment and, consequently, of underdetermined theory change? HSC, I think, provides several illustrations of the inconclusiveness of theory change. However, these illustrations are not as straightforward as one might have hoped and there is a reason why. Suppose, I have found a historical episode where I believe the outcome of theory assessment was inconclusive. Say there was a theory that became accepted and I declare that the acceptance wasn't conclusively prescribed by the then-employed method. Question: how can I ever show this? Recall, that in order to explicate the method employed at the time, I have to study the transitions from one accepted theory to the next during that time period. So whenever I declare that the theory's acceptance wasn't conclusively dictated by the employed method, my opponent can rightly argue that I may be wrong in my explication of the then-employed method. After all, I could only reconstruct the actually employed method by studying the respective changes in the mosaic. All that I know, my opponent may continue, is that the theory did become accepted. Thus, we must conclude that it did somehow satisfy the requirements of the employed method. We have no indication that the assessment of the theory by the then-employed method was inconclusive. My opponent would be absolutely correct in pointing this out: at this stage I cannot tell for sure whether it was an instance of inconclusiveness or whether the actual method wasn't what I thought it was. Indeed, establishing that a particular theory assessment was inconclusive is not a trivial task. If we are to find genuine historical examples of inconclusive theory assessment, we must try more sophisticated approaches. In the next section, I shall discuss one such approach. Namely, I will show that historical instances of inconclusive theory assessment can be detected by means of finding examples of mosaic split. But first we have to understand what *mosaic split* is.

In this section, we have established that the process of scientific change is not strictly deterministic (*the scientific underdeterminism theorem*), for neither method employment nor theory acceptance are always strictly deterministic processes (*the underdetermined method change theorem* and *the underdetermined theory change theorem*). We have also learnt that detecting actual cases of inconclusive theory assessment is not an easy task.

Mosaic Split and Mosaic Merge

Mosaic split is a scientific change that results in one mosaic's turning into two or more mosaics. Imagine a community that initially accepts some theories and employs some methods; in other words, initially, there is one mosaic of theories and methods. Imagine also that (as a result of some events) this initially united community transforms into two different communities with two somewhat different mosaics of theories and methods. In such an instance, we deal with what I define as *mosaic split*:

Mosaic Split ≡
A scientific change where one mosaic transforms into two or more mosaics.

Mosaic split should not be confused with regular *disagreement*, common in any scientific community. Although *mosaic split* is a form of disagreement, not any disagreement among scientists is an instance of mosaic split. The key characteristic of mosaic split is that two different theories are taken to be accepted by two different scientific communities. Namely, there is an instance of mosaic split if and only if each of these communities presents its own theory as the *accepted* one in its articles, encyclopaedias, dictionaries, university lectures etc. Obviously, not every scientific disagreement is like that. Two physicists or even two groups of physicists may disagree on one topic or another. Yet, as long as they take the same theories as accepted ones, there is a regular scientific disagreement. Suppose, for instance, there are two groups of quantum physicists which subscribe to two different quantum theories – say, the so-called Many Worlds theory and GRW theory respectively. Suppose also that the two groups understand that the currently accepted theory is the orthodox quantum mechanics. Consequently, in their university lectures both groups present the orthodox theory as the currently accepted one. Here we have a typical example of scientific disagreement. The members of the two groups may even tell their students that they personally believe there is a better theory available. But as long as they stress that their personal favourite theory is not the currently accepted one, we deal with an instance of regular scientific disagreement. In short, as long as the members of two groups have the same opinion on what the accepted theory is, the scientific community remains united.

Take another example. Imagine a group of physicists circa 1918 who considered general relativity as the best available description of its domain. This view was in disagreement with the position of the vast majority of scientists who believed in the then-accepted version of the Newtonian theory. Yet there was no mosaic split, since both the Newtonians and Einsteinians clearly realised which theory was accepted and which theory was merely a contender. Take Eddington, for instance, who was in that small group of early adherents of general relativity. He had no illusions regarding the status of general relativity, for he knew perfectly well that it wasn't the accepted theory.

Nowadays we have many similar examples. When the so-called creationists question the tenets of the contemporary evolutionary biology, they clearly (albeit reluctantly) admit that the theory they criticize is nevertheless part of the contemporary scientific mosaic. The very fact that they attempt to overthrow the accepted theory is perhaps the best illustration that they understand which theory is the accepted one. Thus, here we deal with another instance of regular scientific disagreement.

Mosaic split, on the other hand, presupposes a very peculiar form of disagreement. To qualify as a mosaic split, it is not enough to have two groups of scientists disagreeing on some issues; the disagreement should concern the very *status* of the theories in question. As long as the debating parties agree that such and such theories are the currently accepted ones, while such and such theories are among contenders, there is no mosaic split, but only a regular scientific disagreement. Take for instance the famous early eighteenth century case of Newtonianism in Britain vs. Cartesianism in France. If we were to go back to the 1730s we would spot at least two distinct scientific communities, with their distinct mosaics. While the curricula of the British universities included the Newtonian natural philosophy, the French universities taught the Cartesian natural philosophy among other things. In short, there is an instance of mosaic split if and only if there are two or more parties that take different theories to be *accepted*. Thus, mosaic split should not be confused with regular disagreement.

The main task of this section is to understand how exactly a mosaic split occurs. The TSC allows for at least two distinct scenarios of mosaic split. On the one hand, a mosaic can split when the requirements of the current method are simultaneously satisfied by two or more competing theories. On the other hand, a mosaic can split when the outcome of theory assessment is inconclusive. While in the former case a mosaic split takes place *necessarily*, in the latter case it is merely *possible*. I shall discuss these two scenarios in turn.

Let us start by recalling that according to *the zeroth law* two mutually incompatible theories cannot be simultaneously accepted in the same mosaic. Now imagine a situation where two mutually incompatible theories simultaneously satisfy the requirements of the currently employed method, i.e. a situation when the assessment outcomes of both theories are conclusive "must be accepted". *The second law* prescribes that they both *must* become accepted. But they obviously cannot become accepted within the same mosaic, for that would violate *the zeroth law*. Thus, we are left with only one logically available option – the two theories must become accepted in two different mosaics and, thus, a mosaic split must take place. In short, when two incompatible theories simultaneously satisfy the current requirements, a splitting of the mosaic becomes inevitable. Here is the deduction of *the necessary mosaic split theorem*:

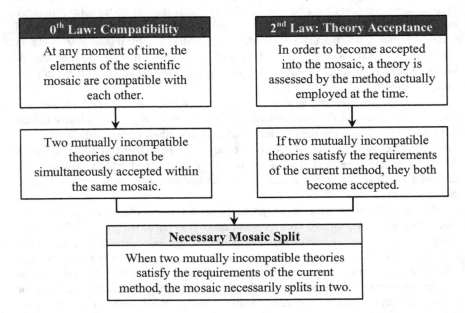

We can tell the same story from the perspective of the scientific community. When two mutually incompatible theories simultaneously satisfy the implicit requirements of the scientific community, members of the community are basically in a position to pick either one. And given that any contender theory always has its champions (if only the authors), there will inevitably be two parties with their different preferences. As a result, the community must inevitably split in two.[52]

The case of necessary mosaic split is a relatively simple one. But it is only one possible scenario of mosaic split. Another possible scenario of mosaic split has to do with *inconclusiveness* of theory assessment outcome. In order to show this I have to consider two hypothetical cases – with one and two contender theories respectively.

Consider first the case with one contender theory. Suppose there is a scientific mosaic with its theories and methods and there is also a contender theory which becomes assessed by the currently employed method. If the assessment outcome is conclusive "accept", the theory necessarily becomes accepted. If the outcome is conclusive "not accept", the theory remains unaccepted. Both of these cases are quite straightforward. But what will happen if the outcome turns out to be inconclusive, i.e. if the assessment by the current method doesn't provide a definitive prescription? When the assessment outcome is inconclusive, there are three possible courses of events. First, the new theory can remain unaccepted; in that case the mosaic will maintain its current state. Second, the new theory can also become

---

[52] Note, however, that I have presented this community-oriented version of the story only for illustrative purposes – it is irrelevant to the actual deduction of *the necessary mosaic split theorem*.

accepted by the whole community; in that case a regular theory change will take place and the new theory will replace the old one. None of these two scenarios is particularly interesting here. However, there is also the third possible course of events. When the outcome of theory assessment is inconclusive, members of the community are free to choose whichever of the two scenarios – they can accept the theory, but they can equally choose to leave it unaccepted. Naturally, there are no guarantees that *all* of them will necessarily choose the same course of action. It is quite conceivable that some will opt for accepting the new theory, whereas the rest will prefer to keep the old theory. In other words, when the assessment of a contender theory yields an inconclusive outcome, the mosaic *may* split in two. Note that here we only deal with a *possibility* of splitting:

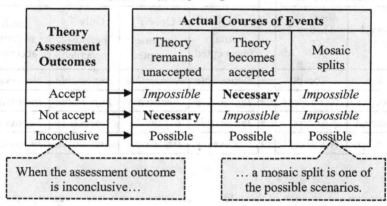

Theory Assessment Outcomes	Actual Courses of Events		
	Theory remains unaccepted	Theory becomes accepted	Mosaic splits
Accept	*Impossible*	**Necessary**	*Impossible*
Not accept	**Necessary**	*Impossible*	*Impossible*
Inconclusive	Possible	Possible	Possible

When the assessment outcome is inconclusive… … a mosaic split is one of the possible scenarios.

The case with two contender theories is more illustrative. When two contender theories undergo assessment by the current method, each assessment can have three possible outcomes. Therefore, there are nine possible combinations of assessment outcomes overall and, in five of these nine combinations, there is an element of inconclusiveness:

Assessment Outcomes	
Theory 1	Theory 2
Accept	Accept
Accept	Not accept
Not accept	Accept
Not accept	Not accept
Accept	Inconclusive
Inconclusive	Accept
Not accept	Inconclusive
Inconclusive	Not accept
Inconclusive	Inconclusive

There is an element of inconclusiveness in these five combinations.

The actual course of events in the first four combinations is relatively straightforward. If the assessment of one theory yields a conclusive "accept" while the assessment of the other yields a conclusive "not accept", then, by the second law, the former becomes accepted while the latter remains unaccepted. When the assessments of both theories yield conclusive "not accept", then both remain unaccepted and the mosaic maintains its current state. Finally, when the assessment yields "accept" for both theories, then both theories become accepted and a mosaic split takes place, as we know from *the necessary mosaic split theorem*:

Assessment Outcomes		Actual Courses of Events			
Theory 1 (T_1)	Theory 2 (T_2)	Both remain unaccepted	Only T_1 becomes accepted	Only T_2 becomes accepted	Both become accepted
Accept	Accept →	*Impossible*	*Impossible*	*Impossible*	**Necessary**
Accept	Not accept →	*Impossible*	**Necessary**	*Impossible*	*Impossible*
Not accept	Accept →	*Impossible*	*Impossible*	**Necessary**	*Impossible*
Not accept	Not accept →	**Necessary**	*Impossible*	*Impossible*	*Impossible*

As we can see, in each of these four cases, there is only one necessary course of events. In other words, when the assessment outcomes of both theories are conclusive, the actual course of events is strictly determined by the assessment outcomes. This is not the case with the other five combinations of assessment outcomes. Let us consider them in turn.

"Accept"/"inconclusive" (2 combination): What can happen when the assessment of one theory yields a conclusive "accept", while the assessment outcome of the other theory is inconclusive? In such a scenario, the former theory must necessarily become accepted, while the latter may or may not become accepted. Therefore, only two courses of events are possible in this case: it is possible that only the former theory will become accepted and it is also possible that both theories will become simultaneously accepted (i.e. a mosaic split may take place).

"Not accept"/"inconclusive" (2 combinations): What can happen when the assessment of one theory yields a conclusive "not accept", while the assessment outcome of the other theory is inconclusive? In such an instance, it is impossible for the former theory to become accepted, while the latter may or may not become accepted. Thus, it is possible that both theories will remain unaccepted as well as it is possible that only the latter theory will become accepted. Finally, the mosaic split is also among the possibilities, since it is conceivable that one part of the community may opt for accepting the latter theory while the other part may prefer to maintain the current state of the mosaic. Disregard for a moment the former theory: it cannot become accepted, since

its assessment yields a conclusive "not accept". With the former theory out of the picture, we are left with the latter theory – the one with an inconclusive assessment outcome. Thus, this case becomes similar to the above-discussed case with only one contender theory: we have a contender with an inconclusive assessment outcome and, consequently, a mosaic split may take place provided that one part of the community decides to opt for the theory while the other part prefers to stick to the existing mosaic. Note that, in this case, a split is not a consequence of the simultaneous acceptance of two mutually incompatible theories.

"Inconclusive"/"inconclusive" (1 combination): Finally, what can happen when the assessment outcomes of both theories are inconclusive? In such a scenario, both theories may or may not become accepted. Thus, it is possible that none of the theories will become accepted, just as it is possible that only one of the two will become accepted. It is also possible that both theories will become simultaneously accepted and, consequently, a mosaic split will take place.

This meticulous discussion of possible scenarios leads to an important conclusion: a mosaic split is possible in those cases where the assessment outcome of at least one contender theory is inconclusive:

Assessment Outcomes		Actual Courses of Events				
Theory 1 (T_1)	Theory 2 (T_2)	Both remain unaccepted	Only T_1 becomes accepted	Only T_2 becomes accepted	Both become accepted	Mosaic splits
Accept	Inconclusive →	*Impossible*	Possible	*Impossible*	Possible	Possible
Inconclusive	Accept →	*Impossible*	*Impossible*	Possible	Possible	Possible
Not accept	Inconclusive →	Possible	*Impossible*	Possible	*Impossible*	Possible
Inconclusive	Not accept →	Possible	Possible	*Impossible*	*Impossible*	Possible
Inconclusive	Inconclusive →	Possible	Possible	Possible	Possible	Possible

A mosaic split is possible in each of the five cases, i.e. it is possible whenever a theory assessment of at least one of the theories is inconclusive.

This is, in essence, *the possible mosaic split theorem*. Here is the deduction of the theorem:

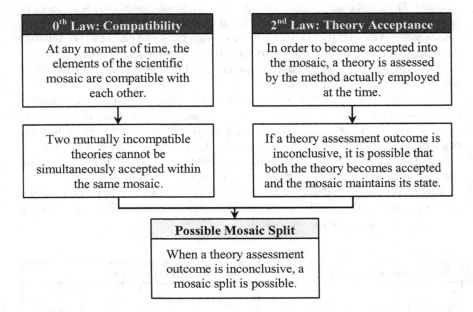

The difference between the two theorems is essential. In the case of a *necessary* mosaic split, the split occurs inevitably – it is the only possible course of events. For that kind of split to take place, there should be two contenders with assessment outcomes "accept". The case of a *possible* mosaic split is different: it may or may not take place. The sufficient condition for this second variety of mosaic split is an element of inconclusiveness in the assessment outcome of at least one of the contender theories.[53]

Before proceeding to historical examples of mosaic split, I must stress one more important point. As I have indicated at the very end of section "Scientific Underdeterminism", if there have been any actual cases of inconclusive theory assessment, they can be detected only indirectly. One way of detecting an inconclusive theory assessment is through studying a particular instance of mosaic split. Unlike inconclusiveness, *mosaic split* is something that is readily detectable. As long as the historical record of a time period is available, it is normally possible to tell whether there was one united mosaic or whether there were several different mosaics. For instance, we are quite confident that in the seventeenth and eighteenth centuries there were many differences between the British and French mosaics.

Now, how exactly can we detect a case of inconclusive theory assessment? First, let us appreciate that a mosaic split *can* but *need not* necessarily be a result of incon-

---

[53] There is an interesting theoretical question that calls for further study: is it possible for a mosaic to split as a result of the employment of two incompatible methods? In other words, does mosaic split always result from *theory* acceptance or can it also result from *method* employment?

clusiveness. When a mosaic split takes place it may be a result of inconclusiveness, but it may also be a case of necessary mosaic split, covered by *the necessary mosaic split theorem*. Thus, not every instance of mosaic split is an indication of a preceding inconclusive theory assessment. There is, however, a species of mosaic split which can only be a result of inconclusiveness. I am referring to the case with one contender theory, where the theory becomes accepted only by one part of the community, with the other part opting for maintaining the current state of the mosaic. If we manage to find such a case, it will be a clear-cut indication of inconclusiveness. I shall call this *the split due to inconclusiveness theorem*:

Split due to Inconclusiveness
When a mosaic split is a result of the acceptance of only *one* theory, it can *only* be a result of inconclusive theory assessment.

Note that a mosaic split can occur differently. It may occur when there are two or more new theories involved, or it may also occur if there is only one new theory. Let us consider both cases in turn.

When a mosaic split is a result of the acceptance of two new theories, it may or may not be a result of inconclusiveness. Indeed, it is possible that the assessment of both theories yielded conclusive "accept" and the split took place in accord with *the necessary mosaic split theorem* without involving any inconclusiveness. But it is also possible that the assessment of at least one of the two theories yielded "inconclusive" and, consequently, the split occurred in accord with *the possible mosaic split theorem*. Therefore, the case of a mosaic split with two new theories doesn't necessarily indicate that there was any inconclusiveness involved:

Possible Mosaic Split	**Necessary Mosaic Split**
When a theory assessment outcome is inconclusive, a mosaic split is possible.	When two mutually incompatible theories satisfy the requirements of the current method, the mosaic necessarily splits in two.

When a mosaic split is a result of the acceptance of two mutually incompatible theories, it may or may not be a result of an inconclusive theory assessment.

Thus, if we are to detect any instances of inconclusive theory assessment, we must refer to the case of a mosaic split that takes place with only one new theory becoming accepted by one part of the community with the other part sticking to the old theory. This scenario is covered by *the possible mosaic split theorem*. We can conclude that when a mosaic split takes place with only one new theory involved, this can only indicate that the outcome of the assessment of that theory was inconclusive:

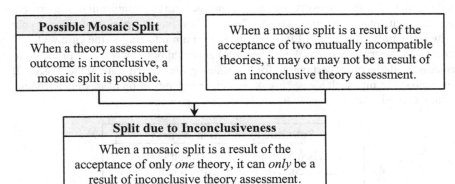

With this theorem at hand, we now know how to detect cases of inconclusive theory assessment. Luckily, HSC provides us with several illustrations of this phenomenon. In particular, I am referring to the transition from the Aristotelian-medieval natural philosophy to the Cartesian and Newtonian natural philosophies in the seventeenth to eighteenth centuries. If we look at the transition from a bird's eye perspective, we will probably notice that the Aristotelian-medieval natural philosophy was taught in the universities across Europe up until the end of the seventeenth century and then, circa 1700, it was replaced by the Cartesian theory in France and the Newtonian theory in Britain. At first glance, it may appear as if it were a clear-cut case of mosaic split. Indeed, it is easy to jump to this conclusion; if we focus merely on respective *natural philosophies*, we will see a transition from one natural philosophy to two different but simultaneously accepted natural philosophies. Consequently, we may end up thinking that this is a case of mosaic split.

However, the actual situation was more complex. For one, the British and French mosaics weren't absolutely identical even before the transition. Naturally, both mosaics included the same propositions of the Aristotelian-medieval natural philosophy, but they differed in their respective accepted *theological* propositions. Recall, for instance, the doctrine of transubstantiation which was accepted in the French mosaic, but wasn't part of the British mosaic. This difference, as we have seen, proved to be crucial.[54] Therefore, 1700 was not exactly the year when the splitting of these two mosaics began; the origins of splitting should be traced back to the mid-sixteenth century English Reformation, which resulted in the rejection from the British mosaic of several propositions of Catholic theology. Namely, it can be traced back to the famous *Book of Common Prayer* of 1549 and the *Acts* of 1558 which basically stated that the doctrine of transubstantiation, to put it mildly, was not the only possible explanation of Real Presence. A century later, in 1662, the same position was reaffirmed and even strengthened in Article 28 of the *Thirty-Nine Articles* which says unequivocally that "Transubstantiation (the change of the substance of

---

[54] Refer to the discussion of the Eucharist episode in section "Contextual Appraisal".

the bread and wine) in the Supper of the Lord cannot be proved from holy Scripture, but is repugnant to the plain teaching of Scripture. It overthrows the nature of a sacrament and has given rise to many superstitions."[55] In short, the British (Anglican) and the French (Catholic) mosaics had been separated at least since the mid-sixteenth century, for the former didn't include the doctrine of transubstantiation. Thus, what happened circa 1700 was merely a widening of the gap between the two mosaics – it doesn't qualify as a proper mosaic split.

What seems to qualify as a proper mosaic split is the acceptance of the Cartesian natural philosophy in Cambridge circa 1680. Let us begin with the available historical data. Prior to the 1680s, the Aristotelian-medieval natural philosophy was taught in schools across Europe, with alternative theories included into the curricula only sporadically. If my understanding is correct, the first university where the Cartesian natural philosophy was accepted and taught on a regular basis was Cambridge. Although the theory had been sporadically taught since the 1660s, it began to be taught systematically only circa 1680.[56] Thus, it is not surprising that when one Cambridge professor Isaac Newton was writing his magnum opus, the main target of his criticism was Descartes's theory, not that of Aristotle. According to the historical data, during the last two decades of the seventeenth century, Cambridge remained the only university where the Cartesian theory was generally accepted. The situation changed circa 1700, when the Cartesian natural philosophy together with its respective modifications by Huygens, Malebranche and others became accepted in France,[57] Holland[58] and Sweden.[59] As for Oxford, it never accepted the Cartesian theory but switched directly to the Newtonian theory circa 1690.[60] In Cambridge, the transition from the Cartesian natural philosophy to that of Newton took place in the 1700s.[61] Most likely, the universities of the Dutch Republic (Leiden and Utrecht) were the first on the Continent to accept the Newtonian theory by 1720.[62] In France and Sweden, the Newtonian theory replaced the Cartesian natural

---

[55] Cummings (ed.) (2011), p. 681.

[56] See Gascoigne (1989), pp. 54–55. As with almost all dates of theory acceptance, a more precise data is required here.

[57] See Brockliss (2003), pp. 45–46, (2006), p. 260.

Vartanian gives a slightly later date: on his reckoning, the Cartesian natural philosophy was included in the curricula of the University of Paris only in the 1710s. See Vartanian (1953), p. 41. However, the rule of thumb suggests that in such matters we have to rely on the more recent scholarship.

Also, McLaughlin mentions that the theory was officially recognized only in the 1720. See his (1979), p. 569. However, it is safe to say that this was merely post factum recognition by the authorities of what had been already apparent: by the time of the official recognition the theory had been systematically taught in Paris for two decades.

[58] See Schmitt (1973), p. 163.

[59] See Frängsmyr (1974), p. 31.

[60] See Gascoigne (1989), p. 146.

[61] See Gascoigne (1989), pp. 145, 155; Turner (1927), pp. 47–49.

[62] Although it must be pointed out that in the Dutch Republic Newton's theory acquired a very special flavor. See Jorink and Maas (eds.) (2012).

philosophy circa 1740.[63] The picture wouldn't be complete if we didn't mention the important theological differences: Catholic theology was accepted in Paris; Anglican theology was accepted in Oxford and Cambridge; in Holland and Sweden the accepted theology was that of Protestantism. Here is a draft timeline:

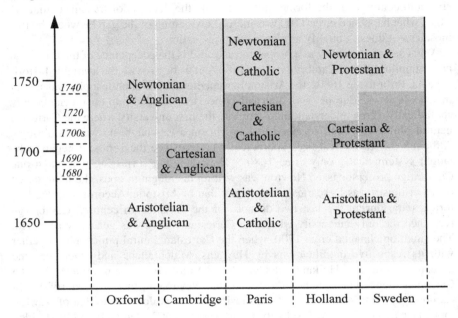

Although the diagram hardly scratches the surface of the colorful seventeenth to eighteenth century landscape, it points to at least two possible candidates of mosaic split. Apparently, there seem to have been a split in the Anglican mosaic of Britain circa 1680, when the Cartesian natural philosophy became accepted in Cambridge, and also probably in the Protestant mosaic sometime by 1720, when the Newtonian theory became accepted in Holland. I shall leave the case of the Dutch Republic for a future study and shall consider rather the more studied case of the acceptance of the Cartesian theory in Cambridge.

If my reading is correct, then this was a typical case of mosaic split: after the acceptance of the Cartesian theory, the mosaic of Cambridge became different from the Aristotelian-Anglican mosaic of other British universities.[64] Note that this mosaic split was caused by the acceptance of only one new theory. Therefore, it could only be a result of an inconclusive theory assessment. At this point, we can only hypothesize as to why exactly the outcome of the assessment of the Cartesian theory was inconclusive.

---

[63] See Aiton (1958), p. 172; Frängsmyr (1974), p. 35.

[64] My reading holds only if the two mosaics had not been split earlier than the 1680s. Whether the Cambridge and Oxford communities shared the same mosaic before 1680s, or whether they had different mosaics is for HSC to establish.

My historical hypothesis is that it had to do with the inconclusiveness of the Aristotelian-medieval method employed at the time, i.e. with the vagueness of the implicit expectations of the community of the time. It is easily seen that the then-employed Aristotelian-medieval method allowed for two distinct scenarios of theory assessment.[65] On the one hand, if a proposition was meant as a theorem, it was only expected to show that it did in fact follow from other accepted propositions. That much would be sufficient for a new theorem to become accepted. This part of the method is straightforward – no ambiguity here. If, on the other hand, a proposition was not meant as a theorem – if it was supposed to be a separate axiom – then it was expected to be intuitively true. But what does it mean to be intuitively true? Nowadays we seem to realize that no proposition can be intuitively true (unless of course it is a tautology) and that intuition, even when "schooled by experience", is not the best advisor in theory assessment.[66] Therefore, a theory could merely *appear* intuitively true to the community of the time. This was the actual expectation of the scientific community in the seventeenth century – the appearance of intuitive truth. One indication of this is the fact that both Descartes and Newton understood the vital necessity of presenting their systems in the axiomatic-deductive form. They also made all possible efforts to show that their axioms – the starting points of their deductions – were beyond any reasonable doubt. They both realized that if their theories are ever to be accepted, their axioms must appear clear to anyone who is knowledgeable enough to understand them. But this is exactly what was expected by the scientific community of the time.

Yet, the requirement of intuitive truth is extremely vague: what appears intuitively true to me need not necessarily appear intuitively true to others. I think, this can explain why the mosaic split of the 1680s took place. The axioms of the Cartesian natural philosophy were meant as self-evident intuitively true propositions. But as with any "intuitive truth", scientists could easily disagree as to whether the axioms were indeed intuitively true. As a result, the outcome of the assessment of the Cartesian theory was "inconclusive". In that situation, a mosaic split was one of the possible courses of events (by *the possible mosaic split theorem*). Of course the mosaic split wasn't inevitable – it was merely one of the possibilities which actualized. This Aristotelian "bring before me intuitive true propositions" requirement was so vague that theory assessment could easily yield an "inconclusive" outcome and, consequently, result in a mosaic split. It is not surprising, therefore, that the British mosaic did actually split in the 1680s when the Cartesian natural philosophy was accepted only in Cambridge.

This was an instance of *possible* mosaic split, but how about *necessary* mosaic split, i.e. a split that occurs when the assessments of two theories simultaneously yield a conclusive "accept"? Have there been any actual cases of necessary mosaic split? Have there been any cases where a split was forced by two contenders conclu-

---

[65] Refer to section "The Third Law: Method Employment" for my explication of the Aristotelian-medieval method.

[66] Bunge (1962) provides a nice discussion of different species of intuitivism and shows why it is doomed.

sively satisfying the requirements of the employed method? The short answer is "none that I know of". There certainly have been cases of mosaic split, but all such cases that I can think of are instances of possible mosaic split, i.e. a split due to inconclusiveness; none of them was a case of necessary mosaic split. Detecting actual historical cases of necessary mosaic split would be an extremely interesting challenge for HSC.[67]

From mosaic split, let us now turn to the opposite transition which would be reasonable to call *mosaic merge*. Here is the definition:

> **Mosaic Merge ≡**
>
> A scientific change where two mosaics
> turn into one united mosaic.

Suppose there are two scientific communities with their respective mosaics. Question: what should happen in order for the two mosaics to become one? If there are two mosaics, it means that there are elements present in one mosaic and absent in the other. To use the language of set theory, these are the elements that constitute the so-called *symmetric difference* of two mosaics:

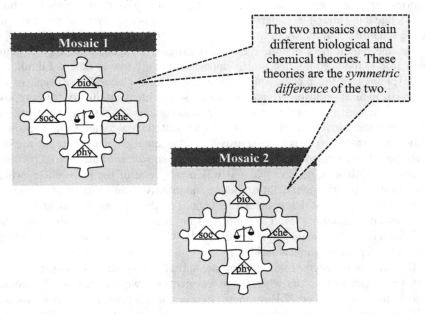

The two mosaics contain different biological and chemical theories. These theories are the *symmetric difference* of the two.

Therefore, in order for the two mosaics to merge into one, these elements should either be rejected in both or accepted in both, so that the differences between the two

---

[67] In case the community expects theories of scientific change to provide some novel predictions, necessary mosaic split can be considered one such prediction of this TSC.

are resolved. At this point, the reader may rightfully ask "is that all you can say about *mosaic merge*?". Indeed this doesn't say much, for to say that mosaics merge when the differences between them are resolved amounts to saying that mosaics merge when they consist of the same elements. But this is just a restatement of the *definition* of the term!

The truth is my attempts to come up with a meaningful theorem about mosaic merge have been fruitless. All the scenarios of mosaic merge that I have been able to think of can be expressed by an extremely trivial proposition that "mosaics merge when they no longer contain different elements". Naturally, I am hesitant to formulate this triviality as a theorem. So at the moment there are no theorems about the process of mosaic merge. This leaves us with a profound maxim: mosaics merge when they, well, merge!

As a possible historical example of mosaic merge it might be tempting to consider the acceptance of the Newtonian theory in the 1740s when the French and Swedish mosaics acquired the same natural philosophy as those of Britain and Holland. However, since the mosaics still contained different theological elements, we must admit that even after the 1740s there remained at least three different mosaics – Catholic, Anglican, and Protestant.[68] Yet, the acceptance of the Newtonian theory did lead to a mosaic merge. Namely, it led to the merging of the Dutch and Swedish mosaics into a unified mosaic with the Newtonian natural philosophy and Protestant theology.[69]

It is well known that, on most of the Continent, the Newtonian theory (together with its eighteenth century modifications) became accepted only after the confirmation of one of its novel predictions. Although, according to popular narratives, the theory was confirmed only in 1758 after the return of Halley's comet,[70] it is safe to say that it was actually confirmed in the period between 1735 and 1740 during the observations of the Earth's shape.

The story goes like this. In 1735, the accepted natural philosophy on most of the continent was the updated version of the Cartesian theory, which assumed that the Earth must be slightly elongated at the poles. The assumption that the Earth is a prolate spheroid was also in accord with the results of the geodesic measurements of Giovanni Domenico Cassini and his son Jacques Cassini announced in 1718. Initially, however, the Earth's prolateness wasn't a consequence of the Cartesian natural philosophy. When Jacques Cassini announced his results, the accepted theory of gravity was a version of Descartes's vortex theory modified by Huygens. According to Huygens's theory, the equilibrium state of any homogenous fluid

---

[68] See the timeline on page 208.

An interesting historical question arises here: when did the three mosaics merge? Possibly, it had to do with the rejection of respective theological propositions. If that is the case, then the task is to find out when theology was exiled from the mosaic.

[69] It is probable that the other protestant mosaics, such as that of German scientific community, also merged with those of Dutch and Swedish mosaics. This can be revealed only by proper historical research.

[70] See, for instance, Cohen (1985), pp. 182–183.

mass, subject to aethereal pressure, was not prolate but oblate spheroid. Thus, in 1718, the prolateness of the Earth announced by Cassini was an anomaly for the accepted Cartesian natural philosophy. In the period between 1720 and 1734 several attempts were made to reconcile the results of Cassinis' measurements with the accepted theory of Huygens. There is no unanimity among the historians as to which reconciliation became actually accepted. On my reckoning, it was the very first reconciliation provided by Mairan in 1720, which absorbed the anomaly by stipulating the Earth's primitive prolateness.[71] In any case, we know for sure that by 1735 the prolate-spheroid Earth was already part of the accepted version of the Cartesian natural philosophy. As for the Newtonian theory (which was a contender at that time), it was predicting that the Earth is slightly flattened at the poles, i.e. that the Earth is an oblate spheroid.

In order to end the controversy, the French *Académie des Sciences* organized two expeditions to Peru (1735–1740) and to Lapland (1736–1737). The latter expedition led by Maupertuis who was accompanied, among others, by Swedish astronomer Anders Celsius, returned to Paris in the summer of 1737. Its results showed that the prediction of Newton's theory was correct.[72] This conclusion was also confirmed by Jacques Cassini's son César-François Cassini de Thury who re-measured the Paris-Perpignan meridian in 1740.[73] As a result, the Newtonian theory replaced the Cartesian theory in all the mosaics where the latter was accepted. In particular, this resulted in the merging of all protestant mosaics where the Newtonian theory became accepted (the Dutch and Swedish, in our timeline).[74]

Let us sum up the main outcome of this section. We have discussed two different types of mosaic split – necessary and possible. The former is inevitable when two mutually incompatible theories conclusively satisfy the requirements of the employed method; in that case, both theories become accepted, which results in a mosaic split (*the necessary mosaic split theorem*). As for the latter, it is possible whenever the assessment outcome of at least one of the contender theories is inconclusive (*the possible mosaic split theorem*). We have also learnt that in order to detect actual historical cases of inconclusiveness we must look for mosaic splits with only one contender theory becoming accepted (*the split due to inconclusiveness theorem*). Finally, for now, there are no theorems regarding the process of mosaic merge.

---

[71] See Terrall (1992), p. 221; Lafuente and Delgado (1984), p. 21. For an alternative view, see Greenberg (1987), p. 293, who seems to be saying that there was no reconciliation up until 1734, when the solution provided by Johann Bernoulli became accepted by the Académie. See also Greenberg (1995).

[72] For a detailed account of the Lapland expedition, see Terrall (2002). For the Peru expedition, see Lafuente and Delgado (1984).

[73] See Terrall (1992), p. 234.

[74] Again, it needs to be emphasized that my historical hypotheses are to be taken with a grain of salt as they are presented only for the purpose of illustrating the theorems of the TSC. Only professional historical research can establish whether my historical hypotheses hold water.

Static and Dynamic Methods

From the days of Aristotle up until Popper and Lakatos, philosophers generally believed that despite all changes in theories at least one element of the scientific mosaic is immune to change. That element was believed to be the scientific method – the transhistorical set of requirements that any acceptable theory was supposed to satisfy in order to become accepted. By the 1960s–70s, however, it was discovered that the requirements that have been normally associated with the scientific method are not immune to change. The idea that the alleged method of science is dynamic and not static was one of the key tenets of both Kuhn's *Structure* and Feyerabend's *Against Method*. Kuhn, Feyerabend and others illustrated this by numerous historical examples. Yet this stunning discovery raised an important question: are *all* methods of science *dynamic* (i.e. transient, changeable), or are there perhaps some more basic methods which are immune to change after all? By the 1980s it had already been accepted by the philosophers of science that many of the requirements employed by the scientific community in theory assessment are dynamic. It had also been understood that our accepted theories somehow shape our requirements.[75] The case of the placebo effect and its consequences for the method of drug testing had already been well absorbed.[76] Nevertheless, many philosophers still hoped that, despite all apparent transitions in methods, there might be after all a set of more fundamental *static* requirements, i.e. requirements that somehow remain fixed in the course of scientific change.

In the late 1980s, the question of existence of static methods became a focal point of the debate between Laudan and Worrall. In his *Science and Values*, Laudan (or the later Laudan to be precise) argued that no method of theory assessment is immune to change. Worrall disagreed by claiming that there are nevertheless some methods which have persisted throughout all changes.[77]

However, a careful analysis reveals that *initially* two different questions were mixed up in the debate. It is one thing to ask whether there are such requirements that *haven't* changed throughout history, and it is another thing to ask whether there are requirements that are, in principle, *immune* to change. While Worrall was interested in the former issue, Laudan was obviously concerned with the latter.[78] Clearly, the two issues do not coincide, since it is possible that there are requirements which have been always employed in theory assessment ever since the days of ancient science but are, nevertheless, not immune to change. Moreover, the two questions pertain to two different domains. The former is an empirical question that must be

---

[75] See, for instance, Newton-Smith (1981), pp. 222, 245–246, 269.

[76] See section "The Third Law: Method Employment" for details.

[77] The key sources in chronological order: Laudan (1984), Worrall (1988), Laudan (1989), Worrall (1989). Unfortunately, the debate has been unfairly overlooked.

[78] See especially Laudan (1989), p. 371, footnote 6; Worrall (1989), p. 376, 384.

tackled by HSC. On the contrary, the latter is a theoretical question and, therefore, it is the latter question that concerns us here:

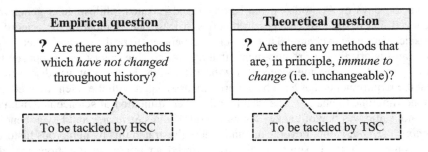

Empirical question	**Theoretical question**
? Are there any methods which *have not changed* throughout history?	**?** Are there any methods that are, in principle, *immune to change* (i.e. unchangeable)?
To be tackled by HSC	To be tackled by TSC

In short, *unchanged* should be distinguished from *unchangeable*: whether there have been any unchanged methods is for empirical (historical) research to establish, while our task as theoreticians is to go further and inquire whether there can be unchangeable methods, i.e. methods that are immune to change. Luckily, by the end of their debate, Worrall and Laudan managed to distinguish between these two questions and came to agree that, although there are unchanged methods, in principle there can be no unchangeable methods.[79] It is this theoretical question that separates *the static method thesis* from *the dynamic method thesis*:

? Can there be any *static* methods, i.e. methods that are *immune* to change?	
Yes	**No**
The Static Method thesis: There can be scientific methods which are immune to change.	**The Dynamic Method thesis:** No method of science can be immune to change.

As I have already indicated, *the static method thesis* was implicit in the conceptions of logical positivists, Popper, Lakatos, the early Laudan and many others. It has its adherents even nowadays. While some authors subscribe to this view explicitly,[80] others seem to have assumed this thesis tacitly.[81] The opposite view expressed in *the*

---

[79] Worrall (1989), p. 387.

[80] See, for instance, Abímbólá (2006), p. 55. Zahar calls it "stability thesis". See Zahar (1982), p. 407. In addition, when Bayesianists argue that scientific reasoning is conducted in accordance with the axioms of probability, they tacitly subscribe to *the static method thesis*. See Howson and Urbach (2006).

[81] As I have noted earlier, the whole contemporary discussion on *the role of novel predictions* has the static method thesis as one of its premises. See section "The Second Law: Theory Acceptance" for details.

dynamic method thesis also has its supporters among philosophers.[82] In addition, this thesis is clearly accepted in HSC.[83] Thus, the debate hasn't run out of steam.

Now, what can the TSC tell us about the existence of *static* methods? To answer this question, I shall start by drawing one important distinction. Logically speaking, we can distinguish between methods of two types. As both Worrall and Laudan agreed, there are methods which are shaped by our accepted theories, i.e. which presuppose something about the world we live in. Such methods are called *substantive*. However, logically speaking, it is also possible to conceive of such methods which do not presuppose anything about the world we live in. Methods of this kind are called *procedural*. As for now, this distinction is somewhat imprecise. If a method is based on some physical theory then it is undoubtedly substantive. If a method isn't based on any theory whatsoever (suppose it is self-evident), then it is clearly procedural. But how about a method which is based exclusively on necessary truths (e.g. tautologies, definitions)? Such a method will certainly be procedural as it wouldn't presuppose any empirical theory. So the key question separating substantive and procedural methods is not whether a method presupposes anything, but what sort of presuppositions it makes. To presuppose *contingent* propositions about the world is one thing, to be based exclusively on *necessary* truths is quite another. Thus, we have to reshape the distinction. A method is said to be *substantive* if it presupposes at least one contingent proposition. The notorious double-blind method is an example of substantive method, since it presupposes several contingent propositions (*unaccounted effects* thesis, *placebo effect* thesis, *experimenter's bias* thesis etc.). Conversely, a method is said to be *procedural* if it doesn't presuppose anything at all or if it presupposes only necessary truths.

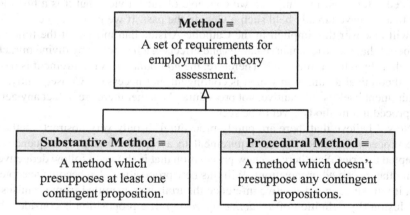

While the Worrall-Laudan debate made it obvious that many methods are substantive, it didn't help to clarify whether there are any procedural methods, i.e. any methods that do not presuppose anything contingent. The methods that

---

[82] See, for instance, Freedman (2009), p. 314.

[83] See Shapin (1996), p. 4.

Worrall had initially thought to be procedural had all turned out to be substantive by the end of the debate. Apparently, what Worrall considered as an example of procedural method in the beginning of the debate was a version of the hypothetico-deductive method that, among other things, required confirmed novel predictions. But as we already know, the hypothetico-deductive method is not procedural. It is substantive, since it evidently presupposes several contingent propositions about the world (e.g. that the world is more complex than it appears in observations, that any phenomenon can be produced by an infinite number of different underlying mechanisms etc.).[84]

Another requirement that Worrall considered procedural was that "non-*ad hoc* accounts should always be preferred to *ad hoc* ones".[85] But it is obvious that this requirement too is not procedural, since it is clearly based on a principle that the objective regularities or the laws of nature – whether strictly deterministic or probabilistic – do not allow for exceptions. We try to avoid *ad hoc* explanations such as "all planets except Mercury obey the laws of Newtonian physics" simply because we tacitly hold that changes in nature occur in a regular fashion and that objects of the same class are governed by the same set of laws which do not allow for exceptions. We can call this tacit assumption *the broad causation principle*, for short. Yet, this tacit assumption is by no means self-evident. It is not a necessary proposition, for, as David Hume has successfully shown some two and a half centuries ago, the opposite view is not self-contradictory. In other words, *the broad causation principle* doesn't necessarily hold in all possible worlds. We can easily conceive of such worlds where some processes obey certain laws, while others are completely irregular. We may even conceive of worlds where the natural regularities are occasionally violated. Moreover, not only can we conceive of such a view, but it is a historical fact that we have already held such a view in the past. If we go back to circa 1500 we will discover that implicit in the Catholic-Aristotelian mosaic of the time was the belief that the strict chain of causes and effects can be broken by divine miracles as well as by human free will.[86] Therefore, the *non-adhocness* requirement is based on a theoretical assumption which is contingent, not necessary. Consequently, the requirement itself is substantive, not procedural. Whether there are in fact any genuine procedural methods is yet to be seen.

Now, I believe that there are purely procedural, if only very abstract, methods. One procedural method is the requirement to accept deductive consequences of accepted theories. Say there is a new proposition that is shown to follow deductively from other accepted propositions. Will this new proposition also become accepted? Yes, it will since in a deductive inference the truth is transmitted from premises to conclusion (by definition of *deductive inference*). If a proposition is considered to follow deductively from some accepted propositions, it is automatically accepted. The requirement to accept deductive consequences of accepted theories is proce-

---

[84] See section "The Third Law: Method Employment" above, pp. 142 ff.

[85] Worrall (1989), p. 386.

[86] Apparently, Aristotle himself wouldn't approve this modification. The fact is, however, that it was this Christianized version of his theory that was part of the medieval mosaic. See Lindberg (2008), pp. 229–230, 248, 252; Kuhn (1957), pp. 92–94.

dural since it only presupposes one necessary proposition – the definition of deductive inference:

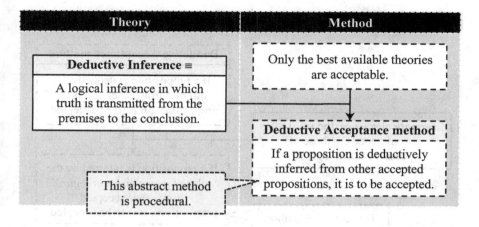

However, it is impossible to say whether one proposition deductively follows from another without specifying a particular logic, i.e. certain rules of inference.[87] Different logics come with different rules of inference: an inference can be valid in one logic and invalid in another. Thus, the specification of the abstract requirement to accept all deductive consequences of accepted theories will depend on what is considered a valid deductive inference and what isn't. This in turn depends on the community's position concerning the applicability of a given rule of inference to different types of propositions of their mosaic. More precisely, the applicability of a given logic to this or that field of science depends on the community's views on the status of the propositions in that field, i.e. whether the accepted propositions of that field are considered strictly true or merely quasi-true. Since those views themselves (fallibilism, infallibilism) are contingent propositions, any specification of the abstract *deductive acceptance method* is going to be a substantive method. I shall explain this.

Imagine a community that holds that all theories in empirical science are fallible. For this community, the accepted propositions of empirical science are only approximately true, quasi-true. This community also knows that two quasi-true propositions can contradict each other.[88] Thus, in order to make sure that contradictions in empirical science do not entail triviality and save their mosaic from explosion, this community tacitly applies some *paraconsistent logic X*. This logic excludes some of those inference rules of classical logic which imply triviality from a contradiction (i.e. *disjunctive syllogism, disjunction introduction*, etc.).[89] Regardless of which of the problematic inference rules is abandoned in this paraconsistent logic X, the important point is that, for this community, a deductive inference is valid only if it

---

[87] This has been pointed out by Rory Harder during the seminar of 2013.

[88] See section "The Zeroth Law: Compatibility" for discussion.

[89] See Burgess (2009), pp. 99–100.

does not lead to triviality. Consequently, by *the third law*, the community's implementation of the abstract method of deductive acceptance will be along these lines:

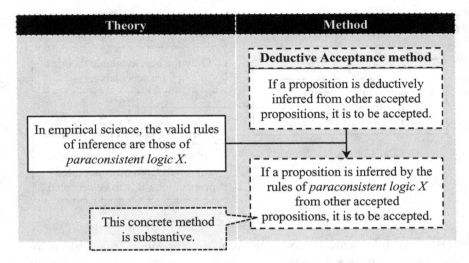

This new specific method is, of course, substantive for it is based on the community's conviction that only paraconsistent logic X is applicable to empirical science. This latter conviction is a contingent proposition, since, among other things, it follows from the belief that theories in empirical science are only quasi-true, i.e. the thesis of *fallibilism*, which is itself a contingent proposition.

In short, while the abstract method of deductive acceptance is procedural, its implementations are substantive methods, for they are based on this or that attitude towards accepted theories. The applicability of certain inference rules to a given field of science depends on the community's views on the status of the propositions in that field. If the accepted propositions of the field are believed to be strictly true, the propositions inferred from the accepted propositions by the rules of classical logic are also to become accepted. Conversely, if the accepted propositions of the field are considered only approximately true, the more cautious inference rules of a paraconsistent logic are applied. It is now the task of HSC to discover which inference laws were employed as criteria of deductive acceptance by different communities at different time periods.[90]

Having clarified the procedural/substantive distinction, we now turn to the main question of this section – that of the existence of *static* methods. It can be noticed that static methods can only *become* employed but they can never *cease* to be employed, for "static" is defined as "immune to change". In other words, if a method is capable of being rejected then it is dynamic, otherwise it is static. Thus, the question of existence of static methods can be reduced to the question of the possibility of *replacement* of a method. If a method can in principle be replaced by another method, it is dynamic; if it cannot, it is static. Thus the question is whether *all* meth-

---

[90] This is another example of an interesting historical question that would probably have never arisen if not for TSC.

ods can be replaced by other methods or whether there are *irreplaceable* methods. The question must be specified for both substantive and procedural methods:

Regarding Substantive methods	Regarding Procedural methods
? Can there be any static (i.e. irreplaceable) *substantive* methods?	**?** Can there be any static (i.e. irreplaceable) *procedural* methods?

The TSC provides answers to both of these questions.

The answer to the first question seems obvious, given our contemporary belief that all contingent propositions are, in principle, fallible. According to the thesis of fallibilism, accepted in the contemporary epistemology, no contingent proposition (i.e. proposition with empirical content) can be demonstratively true. Therefore, since substantive methods are based on fallible contingent propositions, they cannot be immune to change. Imagine a typical mosaic with an accepted theory and a method that implements the constraints imposed by this theory. It is obvious that the method in question is necessarily substantive (by the definition of *substantive method*). Now, suppose that there appears a new theory that manages to satisfy the current requirements and, as a result, replaces the accepted theory in the mosaic. Naturally, this new theory imposes new abstract constraints (by *the third law*). It is conceivable that these new abstract constraints are incompatible with the requirements of the current method. In such an instance, the old method will be replaced by the new one (by *the method rejection theorem*). In short, a rejection of theories can trigger a rejection of the substantive method. This idea has been already implicit in *the synchronism of method rejection theorem*. Thus, there are no guarantees that an employed substantive method will necessarily remain employed *ad infinitum*. Consequently, any substantive method is necessarily changeable, i.e. dynamic. The deduction is quite simple:

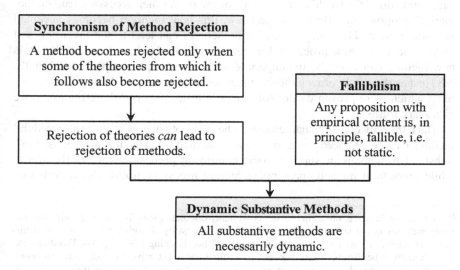

Many of the historical illustrations of method change that I have discussed so far
are also illustrations of this point. The transition from *the controlled trial method* to
the blind trial method and then to *the double-blind trial method* are all examples of
one substantive method replacing another. The transition from *the Aristotelian
method* to *the hypothetico-deductive method* is yet another example.

Those authors who do not draw the procedural/substantive distinction normally
stop here. Indeed, those who do not see any difference between the two are inclined
to jump to a more general conclusion that all methods are dynamic.[91] But such a
conclusion would be premature, for we are yet to consider the case of procedural
methods. The question of static procedural methods is equivalent to the question of
the possibility of replacement of procedural methods. Therefore, the question is
whether one employed procedural method can be replaced by another method. If the
answer turns out to be "yes", then procedural methods are also dynamic. If the
answer turns out be "no", then there can be static procedural methods.

Once more, let us refer to *the method rejection theorem*: a method is rejected
only when other methods that are incompatible with the method in question become
employed. Thus, a replacement of a procedural method by another method would be
possible if the two were incompatible with each other. However, it can be shown
that a procedural method can never be incompatible with any other method – procedural or substantive.

Consider first the case of a procedural method being replaced by another *procedural* method. By definition, procedural methods don't presuppose anything
contingent: they can only presuppose necessary truths. But two necessary truths
cannot be incompatible, since necessary truths (by definition) hold in all possible
worlds. Therefore, two methods based exclusively on necessary truths cannot be
incompatible either; i.e. any two procedural methods are always compatible.
Consequently, by *the method rejection theorem*, one procedural method cannot
replace another procedural method. Consider a new necessarily true mathematical proposition that has been proven to follow from other necessary true mathematical propositions. By the second law, this new theorem becomes accepted
into the mosaic. The acceptance of this theorem can lead to the invention and
employment of a new procedural method based on this new theorem. Yet, this
new method can never be incompatible with other employed procedural methods, just as the newly proven theorem can never be incompatible with those theorems which were proven earlier (of course, insofar as all of these theorems are
necessary truths).

Thus, the only question that remains to be answered here is whether a procedural
method can be replaced by a *substantive* method? Again, the answer is "no".
Substantive methods presuppose some contingent propositions about the world,
while procedural methods presuppose merely necessary truths. But a necessary

---

[91] Hacking, for instance, mixes in one big bowl many disparate methods (some of them even not
being methods in our technical sense) and then goes on asking rhetorically "Where then is this
splendid specific of science, the scientific method?" See Hacking (1996), p. 64. Freedman too
doesn't seem to be distinguishing between procedural and substantive methods and, as a result,
quickly arrives at *the dynamic methods thesis* without touching upon the case of procedural methods. See Freedman (2009), pp. 317, 320.

truth is compatible with any other truth (contingent or necessary). Therefore, no newly accepted theory can be incompatible with an accepted necessary truth. In particular, if we take the principles of pure mathematics to be necessarily true, then it follows that no empirical theory (i.e. physical, chemical, biological, psychological, sociological etc.) can be incompatible with the principles of mathematics. Consequently, a new substantive method can never be incompatible with procedural methods. Take the above-discussed abstract *deductive acceptance method* based on the definition of *deductive inference*: if a proposition is deductively inferred from other accepted propositions, it is to be accepted. It is safe to say that no substantive method can be incompatible with this requirement, for to do so would mean to be incompatible with the definition of *deductive inference*, which is inconceivable. Thus, a procedural method can be replaced neither by substantive nor by procedural methods.

This brings us to the conclusion that all procedural methods are in principle static. Here is the deduction:

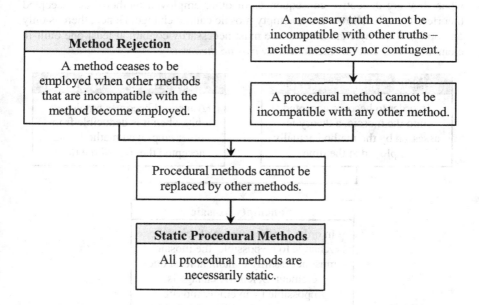

Let us sum up the main outcome of this section. We have established that all substantive methods are necessarily dynamic, whereas all procedural methods are necessarily static. It is worth stressing that this latter result holds regardless of whether our understanding of the nature of logic and mathematics is correct. The only reason why I touched upon logic and mathematics is to provide some illustrations for *the static procedural methods theorem*. The theorem itself doesn't presuppose anything about the necessity of logical or mathematical truths.

Necessary Elements

When we examine *the second* and *the third laws* of scientific change, an interesting question arises. On the one hand, *the second law* suggests that theories become accepted into the mosaic only when they meet the requirements of the method of the time. On the other hand, *the third law* tells us that methods become employed when they are deductive consequences of accepted theories and other methods. Now, if theory acceptance needs at least one employed method and method employment needs at least one accepted theory, then how can the whole enterprise take off the ground?[92]

It follows from *the second* and *third laws* that, in order for the process of scientific change to be possible, any mosaic must contain at least one theory or one method. By *the second law*, no theory can *become* accepted in an empty mosaic, since theory acceptance requires at least one employed method. By *the third law*, no method can *become* employed in an empty mosaic, for methods become employed when they are deductive consequences of other employed methods and accepted theories. Thus, it follows that an empty mosaic cannot change. Hence, there is only one option left: any changing mosaic must necessarily contain at least one built-in element. I shall call this *the nonempty mosaic theorem*:

2nd Law: Theory Acceptance	3rd Law: Method Employment
In order to become accepted into the mosaic, a theory is assessed by the method actually employed at the time.	A method becomes employed only when it is deducible from other employed methods and accepted theories of the time.

Nonempty Mosaic
In order for the process of scientific change to be possible, the mosaic must necessarily contain at least one element. Scientific change is impossible in an empty mosaic.

The history of any community seeking the best possible descriptions of the world must necessarily begin with at least one assumption about the world or one implicit requirement. Our task is to locate this element. What is this element? Is it a theory, or is it a method?

Before we proceed, two questions must be distinguished. It is one thing to ask what elements an *actual* scientific mosaic initially contained. It is a different thing

---

[92] The question was suggested in 2012 by William E. Seager in a private conversation.

to ask what elements must *necessarily* be part of any scientific mosaic. The former question is factual (empirical) and must be tackled by HSC. The latter question, however, is theoretical and must be addressed by TSC:

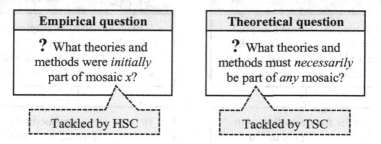

Only special empirical (historical) research can tell what theories and methods were actually present in the initial state of a given mosaic. Thus, only HSC can reconstruct the initial states of the mosaic (or mosaics) of ancient Greeks. Hypothetically speaking, it may turn out that initially they only had one element in their mosaic or, which is more likely, it may turn out that their mosaic initially contained many accepted propositions and a series of implicit requirements. Similarly, only HSC can elucidate what theories and methods were initially present in the ancient Chinese mosaic, or the mosaic of the medieval Islamic scientific community.[93] What concerns us here is the theoretical question: what theories and methods are *necessarily* part of *any* mosaic? This is essentially the question of the *prerequisites* of scientific change: what elements must any mosaic contain in order for the process of scientific change to be possible? The answer is implicit in the laws of scientific change.

We have already established that, any mosaic contains *at least* one element, which is either a theory or a method. But which one is it: is it a theory or is it a method? It is easy to see that if this necessary element of the mosaic were a theory, the process of scientific change would never begin in the first place. Suppose there is a community that accepts only one belief and employs no method whatsoever; this community has no expectations whatsoever. It is obvious that the mosaic of this community will never acquire another element. On the one hand, in order for new theories to become accepted into the mosaic, the mosaic must contain at least one method (by *the second law*). On the other hand, in order for the mosaic to acquire a new method, there must be not only accepted theories, but also at least one other employed method (by *the third law*). Indeed, if we recall the historical examples of *the third law* that we have discussed, we will see that new methods become employed when they are deductive consequences of accepted

---

[93] There are also other interesting *historical* questions concerning the initial state. Was there one initial mosaic or were there many different initial mosaics? Which elements did the initial mosaic(s) contain? When did the mosaic(s) originate? Naturally, all these questions are to be tackled by HSC.

theories and at least one other employed method. Thus, the necessary (indispensable) element cannot be a theory – it must be a method:

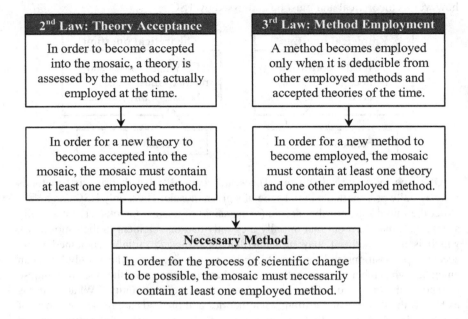

Thus, a community without expectations can never accept a new theory or employ a new method.[94] One method is a must for the whole enterprise of scientific change to take off the ground. But what is this method?

It should be appreciated from the outset that this necessary method cannot be substantive. As defined in section "Static and Dynamic Methods", a method is substantive if it presupposes at least one contingent proposition. Since a substantive method is necessarily based on at least one contingent proposition, it is not a necessary element of any mosaic. Indeed, any substantive method can become employed after the acceptance of those contingent propositions on which it is based. Of course, in some mosaics, substantive methods can also be present from the outset. Moreover, it is quite likely that even the earliest of mosaics tacitly contained some primitive substantive methods (e.g. "trust your senses", or "trust the chieftain"). Again, it is a task of HSC to reconstruct the early states of a given mosaic. Yet, the key theoretical point is that no substantive method is *necessarily* part of any mosaic, for a substantive method can become employed after the acceptance of the theories on which it is based:

---

[94] There is a noteworthy technical detail: logically speaking, *the nonempty mosaic theorem* that we have deduced earlier is a deductive consequence of *the necessary method theorem*: if any mosaic necessarily contains at least one method, then it logically follows that any mosaic contains at least one element.

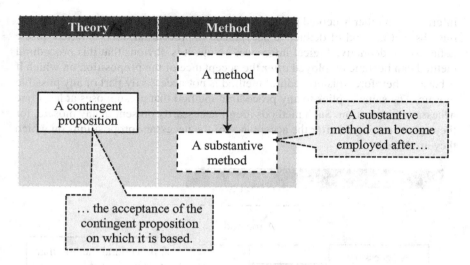

Consider, for example, the requirement of *testability*, according to which a scientific theory must be empirically testable. While many textbooks portray this requirement as one of the prerequisites of science, it is easy to ascertain that it is not a necessary part of any mosaic. To be sure, the requirement of testability appears to be an essential part of our currently employed method. Yet, it is by no means essential for the process of scientific change, i.e. it shouldn't *necessarily* be present in the mosaic from the very beginning. The explanation is simple: the requirement of testability is substantive and, therefore, we can easily conceive of a mosaic where it is not present. It is substantive for it is based, among other things, on such a non-trivial assumption as "observations and experiments are a trustworthy source of knowledge about the world". Thus, the requirement is not a necessarily a part of any mosaic; it can become employed after the acceptance of the assumptions on which it is based. The historical record confirms this conclusion. It is well known that testability hasn't always been among the implicit requirements of the scientific community. For example, it played virtually no role in the Aristotelian-medieval mosaic.[95] The same holds for any substantive method. For instance, the oft-cited requirement of repeatability of experiments is evidently part of our current mosaic, but not of every possible mosaic. Similarly, the requirement to avoid supernatural explanations is implicit in our contemporary mosaic, but it is not a necessary part of any mosaic.[96]

Therefore, the necessary method is not substantive, but procedural, i.e. it doesn't presuppose any contingent propositions. But it is a procedural method of a very special kind in that it cannot presuppose any propositions whatsoever. To appreciate this point, take a procedural method that *does* presuppose some necessary propositions. Let it be the prescription that "if a proposition is deductively

---

[95] See my explication of the Aristotelian-medieval method on pp. 139ff.

[96] For other oft-cited prerequisites of science, see Hansson (2008).

inferred from other accepted propositions, it must also be accepted". As we know, this abstract method of deductive acceptance is procedural, as it is based on the definition of deductive logical inference.[97] Now, it is obvious that this procedural method can become employed *after* the acceptance of the proposition on which it is based. Therefore, this procedural method is not necessarily part of any possible mosaic. The same applies to any procedural method that presupposes at least one necessary proposition. Such methods aren't necessarily present in any mosaic, for they can be employed after the acceptance of the necessary propositions on which they are based:

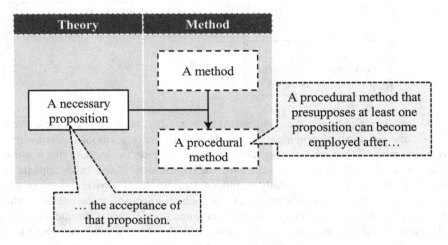

This leaves us with only one option: the method that is necessarily present in any mosaic is not based on *any* propositions. In other words, it must be the most abstract of all methods. Indeed, any concrete method is an implementation of a more abstract method, i.e. any concrete method is a logical consequence of the conjunction of some accepted theories and that abstract method (by *the third law*). Thus, a concrete method can become employed *after* the acceptance of the propositions on which it is based. Therefore, what we are looking for is the most abstract of all possible requirements.

We have come across that requirement on many occasions. I am referring to the most abstract requirement *to accept only the best available theories*. This basic requirement is the most abstract of all, for it does not presuppose any other methods or theories. It is not surprising given that this abstract method is only a restatement of the definition of *acceptance*: this abstract method basically says that a theory is acceptable when it is the best available description of its object, i.e. acceptable. But since this abstract requirement isn't based on any theories, it cannot *become* accepted; it must be *built into* any mosaic from the outset. Conversely, it is safe to

---

[97] See section "Static and Dynamic Methods" for discussion.

say that any other method can be conceived as a deductive consequence of the conjunction of this abstract method and some accepted theories:

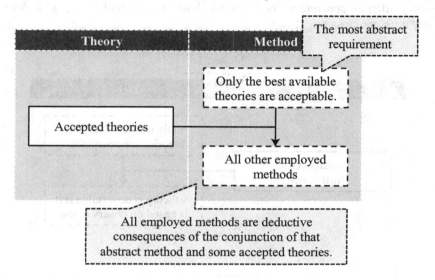

Thus, this abstract method is a prerequisite for the employment of any other method. Hence, it is *the* method that is a necessary part of any mosaic. Of course, this abstract "requirement" is so unrestricting that it doesn't impose any specific constraints upon accepted theories. For instance, it doesn't say how two competing theories are to be compared. It only prescribes to accept the best available theory, but it doesn't specify what makes a theory "the best available description of its object". This prescription may seem virtually useless since to accept a theory is, by definition, to believe that it is the best available description of its object. Yet, as vague and unrestricting as this method is, it nevertheless performs two very important functions. First, it indicates the main goal of the whole scientific enterprise – the acquisition of best available descriptions. Second, being a link between accepted theories and more concrete methods, it allows us to modify our methods as we learn new things about the world, i.e. it allows for concrete methods to become employed as we accept new theories. In short, it is this abstract requirement that makes the process of scientific change possible.

Naturally, this abstract requirement is so vague that virtually any theory has a genuine chance of "satisfying" it. Imagine a community with no initial beliefs whatsoever trying to learn something about the world. In other words, the only initial element of their mosaic is the abstract requirement to accept only the best available theories. Now, suppose they come up with all sorts of hypotheses about the world. Since their method is as inconclusive as it gets, chances are many of the hypotheses will simultaneously "meet their expectations". In such circumstances, different parties will most likely end up accepting different theories, i.e. multiple mosaic splits are virtually inevitable. For example, while some may come to believe that our eyes are trustworthy, others may accept that intuitions (or gut feelings) are the only trustworthy source of knowledge. As a result, the two parties will employ different concrete methods (by *the third law*) and will end up with essentially different mosaics:

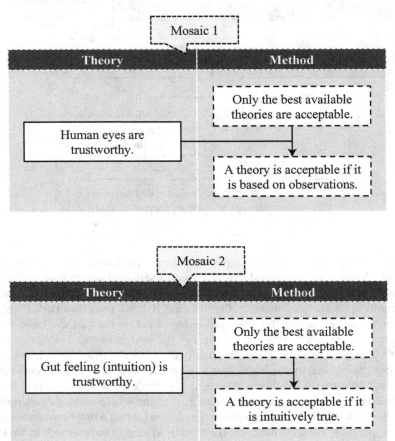

These examples are not altogether fictitious. It is possible that something along these lines happened in ancient Greece, where some schools of philosophy accepted that the senses are, by and large, trustworthy, while other schools held that the senses are unreliable and that the only source of certain knowledge is divine insight (intuition). Thus, the historical fact of the existence of diverse mosaics in the times of Plato and Aristotle shouldn't come as a surprise. Still, it is the task of HSC to study actual initial states of different mosaics and reconstruct their elements (i.e. TSC cannot say how many additional elements this or that actual mosaic initially contained). What is important from the perspective of TSC is that the early steps of any scientific community are, in a sense, random. The initial requirements of any community are inevitably vague and unrestricting. As a result, at early stages, multiple mosaic splits are quite likely.

Let us recap the main outcome of this section. We have deduced that for the process of scientific change to be possible any mosaic must contain at least one element (*the nonempty mosaic theorem*). We have also established that this element is necessarily a method (*the necessary method theorem*). One requirement that any mosaic must contain in order for the process of scientific change to be possible is the abstract requirement to accept the best available descriptions. Although this abstract requirement is extremely vague, it allows for deducing other methods from accepted theories, i.e. it serves as a link between accepted theories and other employed methods.

Sociocultural Factors

Apparently, no other problem appears as troublesome for any theory of scientific change as the problem of the so-called *sociocultural factors*. Can political and economic factor influence the process of theory acceptance and method employment? Do factors such as individual and collective interests influence the process of scientific change? And if they do, does this happen in violation of the laws of scientific change?

Before I proceed to tackling this question, a historical note is in order. Traditionally, the question of the influence of sociocultural factors has been framed in a rather clumsy language of *external* and *internal* factors. Although the traditional distinction between *external* and *internal* factors was rather vague, the idea behind the distinction was simple. On the one hand, there is the world of propositions, which includes our general propositions (e.g. theories of physics or biology) and singular propositions (e.g. results of observations and experiments). On the other hand, there is the society with its economics, politics, culture, personal and collective interests, religions, etc. Everything that has to do with the former was called *internal*, while everything that has to do with the latter was called *external*. This absolute distinction was employed by logical positivists, Popper and others who held that *external* factors such as economics, politics, or religion can only influence the process of theory *construction*, while playing virtually no role in theory *appraisal*.[98]

In his *History of Science and its Rational Reconstructions*, Lakatos attempted to relativize the distinction between *external* and *internal*. In Lakatos's view, the distinction between internal and external factors depends on the methodology that we accept. Consider, for example, the historical phenomenon of adherence to a "refuted" theory (i.e. a theory that is surrounded by anomalies). According to Lakatos, this phenomenon can turn out to be either internal of external, depending on a methodology. Thus, in the falsificationist methodology, the fact that scientists do not rush to reject theories with anomalies is very strange and is relegated to external history; for the falsificationist, the reluctance to reject theories with anoma-

---

[98] See section "Construction and Appraisal" of *Part I* for discussion.

lies can only be explained as a result of political, economic, religious or other factors, but not by the internal logic of science. In contrast, in Lakatos's methodology of scientific research programmes, this phenomenon makes perfect sense and becomes part of the internal history of science; for Lakatos, it is rational for scientists to stick to the theory despite all the anomalies unless there is a better theory on the market. Similarly, only an accepted methodology, according to Lakatos, can tell us whether the requirement of consistency with currently accepted political views is internal or external. Therefore, the demarcation line between *internal* and *external* becomes relative to the methodology that we accept – what counts as external in one methodology may become internal in another.[99]

Now, which of these conceptions should we use here? On the one hand, we cannot stick to the traditional (absolute) distinction between *external* and *internal*, since Lakatos is clearly correct when insisting that only the study of the process of scientific change can reveal which factors are internal to the process and which are external to it. This cannot be given a priori and, thus, it will depend on a theory of scientific change that we accept. On the other hand, Lakatos's own relative distinction is also far from ideal, since it seems to be begging the question. Suppose, we defined *internal factors* as those permitted by the method employed at the time and *external factors* as those not permitted by the method of the time. If we were to proceed in this direction, the question of the influence of external factors would become vacuous. Indeed, we were to define *external* factors as something not covered by the method of the time then we would have to accept that these factors cannot affect the process of theory acceptance (without violating *the second law*). In other words, we would arrive at a vacuous conclusion: the external factors do not affect the process of scientific change because that's precisely how *external factors* are defined. We would arrive at a "solution" to the problem which follows from the definition of *external factors* and, thus, is purely tautologous – external factors do not affect the process, because external factors, by definition, are those factors that do not affect the process.[100] In short, the language of *external* and *internal* doesn't really help us to get to the core of the problem we are interested in.

That is the reason why I will refrain from using the ambiguous and historically loaded concepts of *external* and *internal* and will rather formulate the question in a more neutral language: can sociocultural factors such as individual and group interests, power, religion, politics, economics etc. affect the process of scientific change? If so, under what conditions can they affect the process?

It should be noted from the outset that our TSC can only shed light on the role of sociocultural factors in theory *acceptance* and method *employment*. While theory *construction* appears to be deeply affected by sociocultural factors, this TSC cannot establish whether that is really the case. What we are concerned with here is whether changes in the *mosaic* can be affected by sociocultural factors. This is what constitutes the heart of the matter.

---

[99] See Lakatos (1971), pp. 102, 114, 118–121. Naturally, for Lakatos, *methodology* is a twofold normative-descriptive discipline. See section "Descriptive and Normative".

[100] This is in tune with Garber's criticism of Lakatos's approach. See Garber (1986), p. 98.

Since the process of scientific change involves two types of elements, the question must be specified for both the process of theory acceptance and the process of method employment:

Regarding Theory Acceptance	Regarding Method Employment
? Can sociocultural factors affect the process of *theory acceptance* and, if so, under what conditions can they affect the process?	**?** Can sociocultural factors affect the process of *method employment* and, if so, under what conditions can they affect the process?

Here, I address only the former question, leaving the latter for another occasion.[101]

As we shall see, there are two distinct scenarios of how sociocultural factors can affect the process of scientific change. In the most obvious scenario, sociocultural factors affect the process of scientific change in *violation* of the laws of scientific change. It is these cases, which we normally qualify as unscientific. However, there is also the second scenario where sociocultural factors affect the process of scientific change *in full accord* with the laws of scientific change. Let us begin with the latter scenario.

When we refer to *the second law*, it becomes apparent that sociocultural factors can play part in the process of theory acceptance. In particular, it follows from *the second law* that something can affect a theory's acceptance only insofar as it is permitted by the method employed at the time. If, for instance, the current method prescribes that theories are to be judged by their novel predictions, then confirmed novel predictions will be instrumental for the process of theory acceptance. But if the method prescribes that only intuitively true theories are acceptable, then the community's intuitions will obviously affect the process of theory change. Similarly, if the method of the time ascribes an important role to the position of the dictator or the ruling party then, naturally, the process of theory acceptance will be influenced by the interests of the dictator or the ruling party. In short, it follows from *the second law* that sociocultural factors can affect a theory's acceptance insofar as their influence is permitted by the method employed at the time:

2nd Law: Theory Acceptance	Sociocultural Factors in Theory Acceptance
Theories become accepted only when they satisfy the requirements of the methods actually employed at the time.	Sociocultural factors can affect the process of theory acceptance insofar as it is permitted by the method employed at the time.

---

[101] The question of the role of sociocultural factors in *method* employment is extremely interesting and calls for separate research. At this point we can only notice that there is a good chance that they do play role in method employment; this seems to be suggested by *the underdetermined method change theorem*. Obviously, this needs to be carefully studied.

To illustrate this theorem, let us first consider a case where individual desires and fancies affect the process of theory acceptance. Imagine a hypothetical mosaic which includes an accepted belief that the infallible High Priest always grasps the true essence of things. Naturally, if this proposition is accepted by the community, then the community will also readily accept as infallibly true each and every proposition suggested by the High Priest. This situation is covered by *the third law*: the requirement that a proposition is acceptable if it is proposed by the High Priest is a deductive consequence of the community's belief in the absolute infallibility of their High Priest:

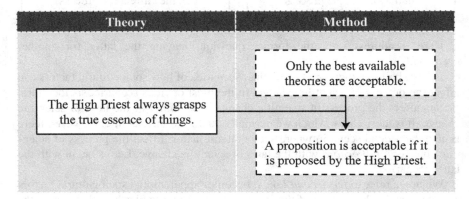

Now, it is obvious that under such circumstances the so-called High Priest can easily manipulate the content of the mosaic. The interests and whims of this High Priest can seriously affect the mosaic of the community. Any nonsense presented by the High Priest will be readily accepted by the community for that is exactly what their method prescribes (i.e. that's what their implicit expectations are). Importantly, in this hypothetical scenario, the laws of scientific change won't be violated. The acceptance of any products of the High Priest's imagination will be in perfect accord with *the second law*. Indeed, *the second law* states that a proposition becomes accepted only if its acceptance is permitted by the method (implicit expectations) of the time. But in this hypothetical case, the expectations of the community are such that they will readily accept everything their High Priest tells them. So the whole process is perfectly conceivable given the laws of scientific change.

Let us now consider a similar scenario when the process of scientific change is affected by *group* interests. Imagine a hypothetical community that believes that God created the world in such a way that its main purpose is to promote the interests of their group. This community considers itself chosen by God in the sense that everything in the world is created to serve the interests of their community. Consequently, this hypothetical community holds that the laws of nature cannot possibly conflict with their interests. Now, by *the third law*, the expectations of this community deductively follow from their accepted beliefs. In particular, this community will never accept a theory, no matter how ingenious or well tested, if it conflicts with their interests. Indeed, those who believe that the whole world was created to promote their interests will only accept theories which are consistent with their interests:

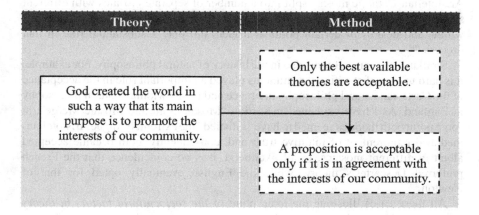

Theory	Method
God created the world in such a way that its main purpose is to promote the interests of our community.	Only the best available theories are acceptable. A proposition is acceptable only if it is in agreement with the interests of our community.

Clearly, in this hypothetical scenario, the community's mosaic will be hugely influenced by the community's interests. Yet, it is easy to see that it will not violate the laws of scientific change; the whole process will be in accord with *the sociocultural factors in theory acceptance theorem*.

Sociocultural factors can also influence the process of theory acceptance when a theory assessment outcome is *inconclusive*. As I have explained in section "Mosaic Split and Mosaic Merge", when the outcome of a theory's assessment is inconclusive there are three possible courses of action: (1) the theory may become accepted, (2) it may remain unaccepted, and (3) a mosaic split may occur. Importantly, the laws of scientific change don't determine which of these three possibilities will actualize if a theory assessment outcome is inconclusive. In such circumstances, individual and group interests may significantly affect the process. In fact, HSC knows many cases when individual and social interests did affect the process. It is safe to say that many of those cases were due to the inconclusiveness of theory assessment.

Take, for instance, the proliferation of philosophical schools in ancient Greece, where each school seemed to have its own distinct mosaic of accepted theories. This situation is covered by *the possible mosaic split theorem*: these multiple mosaic splits were due to the vagueness of the expectations of ancient Greeks regarding philosophical conceptions, i.e. due to the inconclusiveness of their assessment of competing philosophical conceptions. Naturally, in such circumstances, individual and collective sympathies had to play a serious role in deciding the fates of different philosophical conceptions.[102]

---

[102] There is an open historical question concerning the initial implicit expectations of the ancient Greek community (if there ever was *one united* ancient Greek community). Even if it turned out

Analogously, the sixteenth to seventeenth-century Europe witnessed an outpouring of many competing theological systems, several of which became simultaneously accepted by different communities. Consequently, the previously united Aristotelian-Catholic mosaic split into a number of separate mosaics with different systems of accepted theological propositions.[103] It is well known that individual whims and desires of certain political leaders did play a substantial role in that process.[104]

Similar cases can be found also in the history of natural philosophy. For example, it is hard to deny that national sentiments played an important role in the acceptance of the Cartesian natural philosophy in France and the Newtonian natural philosophy in England. As I have explained in section "Mosaic Split and Mosaic Merge", at some point both theories seemed to have "satisfied" the extremely vague Aristotelian-medieval requirement of intuitive truth and, consequently, both became accepted albeit in different mosaics. In all likelihood, it is no coincidence that the French preferred Descartes's theory, while the English eventually opted for that of Newton.[105]

All these cases illustrate the main point of *the sociocultural factors in theory acceptance theorem*: individual and collective whims and desires can affect the process of theory acceptance only when it is permitted by the employed method. What is important is that these cases do not violate the laws of scientific change; sociocultural factors enter the scene in accord with the laws of scientific change.

It is worth repeating that the discussed episodes must not be confused with cases when sociocultural factors simply *violate* the laws of scientific change. Suppose there is an accepted theory, some parts of which are only preserved in a hefty manuscript, and none of the living members of the community knows them by heart. Suppose also that there exists only one copy of this manuscript, since the members of the community are not allowed to copy the text. It is not difficult to imagine what would happen to that mosaic, if the manuscript were stolen or destroyed. Although the community would attempt to restore the mosaic by memory, chances are many parts of the mosaic would be lost forever. This would effectively violate *the law of scientific inertia*, according to which the elements of the mosaic remain in the mosaic unless replaced by other elements. Alternatively, imagine a scientific community all members of which were assassinated. Obviously, the results would be disastrous for the mosaic – the mosaic would simply cease to exist. Again, technically speaking, that would violate *the first law*, for the accepted theories would cease to be accepted without being replaced by any other theory.

These examples are not completely fictitious. For instance, there have been many episodes when governments tried to impose their views on the scientific community and, in some cases, they have succeeded in altering the mosaic of accepted theories

---

that their initial requirements were extremely vague, it would still be interesting to explicate those requirements and determine the degree of their vagueness.

[103] See section "Mosaic Split and Mosaic Merge" for discussion.

[104] See Bernard (2005).

[105] See section "Mosaic Split and Mosaic Merge".

by forcing their own beliefs on the community. History knows many unfortunate cases where members of the community were systematically questioned by their governments and those disputing the position of the rulers were physically eliminated. Recall the infamous case of *Lysenkoism* in the Soviet Union circa 1940, when genetics was declared a "bourgeois pseudoscience" by the Stalin regime and many geneticists were executed or sent to labor camps.[106] Technically speaking, that was a violation of the laws of scientific change.

This conclusion shouldn't come as a surprise, for as I have indicated in section "The Argument from Nothing Permanent", the laws of scientific change are *local*, i.e. they only hold under certain social conditions, such as the existence of a community of scientists relatively independent from the government, sufficient funding, means for publishing and transmitting of knowledge etc. When those conditions are not present, we no longer deal with science. In this sense, the laws of scientific change are similar to any other non-fundamental law. The laws of fundamental physics aside, all other laws are inevitably local and can, in principle, be violated. For instance, the laws of human psychology only hold insofar as certain conditions are met: the brain is not damaged, neurons are not controlled by computer chips etc. This is true for all laws of social sciences; they hold only insofar as there are interacting human beings whose behaviour is not altered by respective physical, biochemical, or social conditions. Take, for instance, the so-called *Engel's law*, an empirical law in economics which states that, as income rises, the proportion of income spent on food falls (even if actual expenditure on food rises). Now, it is obvious that this is true only in "normal" circumstances. It is not difficult to conceive of a scenario when this law is violated (for example, by a dystopian yet conceivable government regulation stipulating that a certain fixed percentage of the household income must always be spent on food). Thus, there is nothing exceptional in the fact that the laws of TSC can, in principle, be violated, since any non-fundamental law can be violated.

Importantly, such violations shouldn't be confused with cases when sociocultural influences are fully in accord with the laws of scientific change. It is one thing when the sociocultural factors affect the process because the High Priest or the ruling elite are actually believed to be a source of trustworthy knowledge. It is another thing when the members of the community are not expecting any words of wisdom from their ruling elite, but the elite nevertheless manages to force its opinion on the community by physically eliminating the disloyal. While in the former case the influence of sociocultural factors does not violate the laws of scientific change, in the latter case the laws are violated and we no longer deal with science proper. *The sociocultural factors in theory acceptance theorem* covers only the influences of the former type, while the latter is to be studied by sociology.

In this section, we established that the laws of scientific change allow for the influence of sociocultural factors. In particular, we deduced *the sociocultural factors in theory acceptance theorem* which states that sociocultural factors such as individual and collective desires can affect the process of theory acceptance when it

---

[106] See Krementsov (1997).

is permitted by the method employed at the time. It may be permitted *directly* (as in the hypothetical cases of the High Priest or the chosen group) or *indirectly* as a result of the vagueness of the method (as in the case of the proliferation of philosophical schools or in the episode of Descartes vs. Newton). Finally, it remains to be studied whether sociocultural factors also play a role in the process of method employment.

The Role of Methodology

The role of *methods* in the process of scientific change should be clear by now – we have already discussed several laws and theorems that concern methods. But we are yet to devise a single theorem that would explicitly concern the role of *methodologies* in the process of scientific change. Based on what we have discussed so far, one could have an impression that methodologies are not capable of affecting our actual implicit requirements (our methods). We know from *the second law* (or even from the respective definitions of *method* and *methodology*) that it is methods and not methodologies that are employed in actual theory assessment. As for methodologies, their role in the process of scientific change remains a mystery. Is it possible for methodologies to play a more *active* role in scientific change? Namely, are methodologies capable of affecting employed methods at all? If so, under what condition can methodologies affect employed methods and, consequently, play a serious part in the process of theory assessment?

Consider the following hypothetical case. Imagine a methodologist who devised some methodology – a set of explicitly formulated rules for assessing theories in some discipline. Naturally, by proposing this methodology, she was hoping that it could one day affect the actual theory assessment: she was hoping that the requirements that she stated in her methodology would eventually become the requirements of the actually employed method. Why else would she bother devising a new methodology? Question: how can she transform her methodology into an actually employed method? In other words, how can her openly formulated criteria become the implicit expectations of the community? Is it even possible for her methodology to influence the implicit expectations of the community at all? It is worth stressing that what I am concerned with here is not how a methodology, as a set of normative propositions, can become openly prescribed by the community and find its place in textbooks and encyclopaedias. This is another open question that needs to be addressed separately. What I am asking here is how, if at all, it is possible for a methodology to shape the actually employed method, the implicit expectations of the scientific community. The answer to this question is far from trivial. To that end, we need to refer to *the third law*.[107]

---

[107] There is also the open question of the status of *normative* propositions (including those of methodology) in the mosaic. Can *normative* propositions such as those of methodology or ethics be part

It follows from *the third law* that methodologies can shape employed methods, but only in very special circumstances. Consider the two basic scenarios of method employment endorsed by *the third law*. In the first scenario, where a method is strictly determined by the accepted theories, no methodology can alter anything. In fact, when a requirement is an immediate logical consequence of some accepted theory, the only way to reject that requirement is by rejecting the theory from which it follows (recall *the method rejection theorem*). Take, for instance, our abstract requirement that a counted number of living cells is acceptable only if it is obtained by an "aided" eye. As we know this abstract requirement strictly follows from our knowledge that the unaided human eye is unsuited to see such minute objects as molecules or living cells. Question: how can this abstract requirement ever be altered? Obviously, as long as we accept that the unaided human eye has these limitations, there is nothing we can do with this abstract requirement. The only way to alter it is by rejecting our theory about the limitations of the unaided eye. But as long as the theory is unchallenged, so is the method that follows from it (by *the synchronism of method rejection theorem*). No methodology, however ingeniously formulated, can change anything in these circumstances:

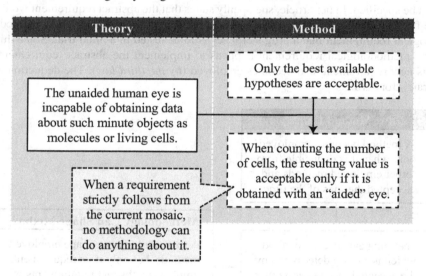

In this scenario, the role of methodology is limited to elucidating our actual expectations; the best an openly stated methodology can hope for in this case is to match our actual expectations accurately.

There is however the second scenario of method employment which gives methodologies a chance to actively influence the process. We have already discussed that a concrete method, which implements abstract requirements of some other method, is not strictly determined by accepted theories. To be sure, the concrete method does

---

of the mosaic? If so, how do they become prescribed, i.e. what logic governs their entrance into and rejection from the mosaic?

follow deductively from accepted theories, but normally this can be shown only *after* the method is devised. The reason is that the concrete method is not the only possible consequence of the theories we accept: each set of abstract requirements can be implemented in many different ways. Recall that *the blind trial method* was not the only possible implementation of the abstract requirement to take into account the possibility of placebo effect.[108]

This is where a methodology can play a decisive role in method employment. To be more precise, a methodology may affect method employment only when its requirements are, simultaneously, implementations of some abstract requirements of other employed methods. Thus, if our methodologist wishes to succeed, she must ensure that the requirements of her methodology specify some more abstract requirements of other employed methods. If the requirements of her methodology indeed manage to implement some of the currently employed abstract requirements, then, by *the third law*, they will become employed in actual theory assessment. For instance, once we accept that there is the possibility of placebo effect, the abstract requirement to take this effect into account becomes immediately employed. Now, suppose that the methodologist writes an article on how this abstract requirement can be specified. In her article, she openly states that the abstract requirement would be fulfilled if we were to perform a blind trial. Thus, she proposes a new methodological requirement that a drug's efficacy is to be tested in a blind trial. Since this new methodological requirement happens to implement the abstract requirement, this new requirement also becomes employed (*by the third law*). The deduction is straightforward:

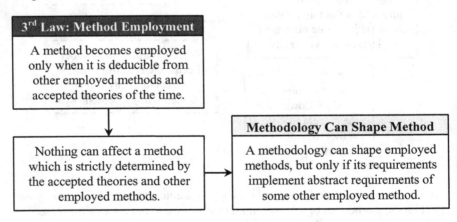

This is the only case where a methodology, a product of human creativity, can shape employed methods (implicit expectations of the community). This is where real innovation in methods is possible. In all other cases, the employed methods are strictly determined by the theories we accept.

---

[108] See section "The Third Law: Method Employment" for discussion.

There is good reason to think that methodologies have actually shaped our employed methods on many occasions. Take our prime example – the case of drug testing methods. It is safe to say that *the double-blind trial method* was first devised as a methodology, as a set of explicitly stated rules, and only *after* that did it become actually employed as a method. In the extreme, it is possible for a methodology to shape our actual expectations almost down to the last detail. Nowadays, we are witnessing such a case: the implicit requirements that we employ in drug testing virtually coincide with the so-called *Good Clinical Practice* (GCP) guidance, a set of openly formulated requirements for clinical trials issued by International Conference on Harmonisation (ICH).[109] In other words, our method of drug testing is almost wholly shaped by our methodology of drug testing. But such extreme cases are few and far between. In the majority of other fields of science, the official methodologies (if any) seem to have only vague resemblance with the actual expectations of the community.[110]

An analogy may be drawn with technology. When devising a plan for a new bridge, we may hope that one day it will turn into an actual bridge. But to give such an outcome a possibility of success, we must ensure that the plan is viable in light of our current knowledge about the world and level of our technology. Only in that case can the plan be employed in building a real bridge. Similarly, a methodology may turn into an actual method but for that, its requirements must be in accord with our accepted theories and methods.

Let us sum up the outcome of this section. Methodologies can affect the methods we employ only in certain circumstances: namely, only when their requirements are implementations of some abstract requirements of other employed methods (*the methodology can shape method theorem*). On the contrary, when an employed method strictly follows from the accepted theories, no methodology can affect the process.

---

[109] For GCP, see Kolman et al. (1998).

[110] Of course, only professional historical research can reveal the distance between the methodology of a field and the method actually employed in that field. This is yet another question which would most likely remain unasked if there were no TSC.

Conclusion

If the TSC had to be summarized in one sentence, it would probably go along these lines: "we stick to our accepted theories until they are replaced by some new theories that satisfy the requirements of our employed methods and then the acceptance of these new theories often leads to changes in our employed methods". What follows is a more elaborate summary.

Although summarizing a full-fledged theory is a virtually impossible mission, some of its main principles are nevertheless worth restating. At the outset I formulated a question: are there any general laws governing the process of scientific change and, if so, how is it possible to unearth these laws? I have tried to answer this question in two steps.

I started by addressing the *metatheoretical* issues of the scope, possibility, and assessment of general theory of scientific change (TSC). Having defined TSC as a theory of transitions within the scientific mosaic, I gradually clarified its scope by showing that TSC is a *descriptive* theory, that it concerns *acceptance* of theories and *employment* of methods by the *scientific community* and that ideally it should account for each and every transition in the mosaic regardless of its scale, time period, or field of inquiry. Once the scope of TSC was outlined, I proceeded to discuss the common arguments against the possibility of general TSC. As we could see, none of the common arguments threatened the prospects of TSC. Then I moved to the issue of the assessment of TSC which concluded *Part I*.

The second step was to build an actual theory of scientific change, which was exactly what I was trying to accomplish in *Part II*. I first postulated four *axioms*, the four laws of scientific change which essentially state the following. Any given state in the process of scientific change is characterized by mutual compatibility of the elements of the mosaic (*the zeroth law*). The process is also characterized by certain inertia as the mosaic of accepted theories and employed methods normally tends to maintain its state (*the first law*). Theories become accepted into the mosaic only when they meet the implicit requirements of the time (*the second law*). As for the requirements themselves, they become employed only if they happen to be deductive consequences of the mosaic of the time (*the third law*).

© Springer International Publishing Switzerland 2015
H. Barseghyan, *The Laws of Scientific Change*, DOI 10.1007/978-3-319-17596-6

From these laws I deduced a number of interesting *theorems*. Specifically, it follows from the laws that the process of theory assessment is an assessment of a proposed modification of the mosaic by the method employed at the time (*the contextual appraisal theorem*). In addition, it became clear that method employment isn't necessarily synchronous with the process of theory acceptance (*the asynchronism of method employment theorem*), whereas method rejection is necessarily synchronous with the rejection of some of the theories on which the method in question is based (*the synchronism of method rejection theorem*). Another deductive consequence of the laws is that neither theory change nor method change are necessarily deterministic processes (*the scientific underdeterminism theorem*). There are also several theorems which explain the mechanism of mosaic split (*the necessary mosaic split theorem, the possible mosaic split theorem, the split due to inconclusiveness theorem*). Finally, it turns out that all substantive methods are necessarily changeable, whereas all procedural methods are necessarily static (*the dynamic substantive methods theorem* and *the static procedural methods theorem*). Overall, at the moment, there are more than 20 deduced theorems.[1]

By proposing this TSC, it was my goal to build on the success of the history of science and unearth the general mechanism of scientific change. The last several decades have witnessed a growing refinement of our historical narratives – more and more details have been added to our knowledge of historical episodes. Compared to the founding fathers of HPS, we are nowadays in a much advantageous position, as both the quality and the quantity of our historical narratives have rocketed since the 1970s. Thus it would be unwise not to take advantage of this situation and not to develop a general TSC. As the proliferation of studies in natural history eventually led to the development of general theories of biology and geology, so the increasing refinement of narratives in the field of history of science has made it possible to detect the general patterns of the process. My task was to study these patterns and create a non-whiggish general TSC that would do justice to historical episodes. I cannot say whether the TSC constructed in this book will ever become accepted. Only the future can tell whether it does or doesn't meet the implicit expectations of the scientific community. Of course, it may turn out to be not as good as I might have hoped, but one has to start somewhere.

The acceptance of this TSC would have significant practical consequences for HSC, as the two would enter into a fruitful cooperation similar to that between theoretical physics and cosmology. Normally, when the cosmologist studies the previous states of the physical universe, she employs the laws of accepted physical theory for reconstructing these states and explaining transitions from one state to the next. Similarly, if this TSC became accepted, HSC would be able to employ the laws of the theory in order to reconstruct scientific mosaics of the past and explain changes in these mosaics. Imagine a historian who studies the process of acceptance of some theory by some community at time *t*. With the laws of TSC at hand, the historian would know that there is no need to ask whether the acceptance of the theory was in

---

[1] It is also clear that there are many questions that this TSC leaves unanswered. See "Appendix: Some Open Questions".

accord with the requirements of the method employed at time t, for clearly it had to be (by *the second law*). Instead, the historian would try to reconstruct the requirements of that employed method, i.e. the requirements that allowed for the theory to become accepted. Having explicated the method employed at the time, the historian would then recall that those requirements had to be deductive consequences of the theories accepted at the time (by *the third law*). So the historian would try to locate those accepted theories from which the requirements of the time deductively followed. Also, the historian would know that the previously accepted theory was rejected because it was incompatible with the new accepted theory (by *the theory rejection theorem*) and thus would try to reconstruct the criteria of compatibility employed by the community. It is safe to say that all the axioms and theorems of TSC would have similar practical implications. When properly applied, the TSC could lead to the discovery of hitherto unobserved historical facts and even whole classes of phenomena, such as, for instance, the phenomenon of *necessary mosaic split* predicted by this TSC.

Therefore, it is my suggestion that we agree on what we have established so far, in order to know where we currently stand and then to move on towards a better and improved understanding of the mechanism of scientific change – towards a better theory of scientific change.

Appendix: Some Open Questions

Metatheoretical Questions

- What is *scientific community*? Does "scientific community" need a definition other than "the bearer of a scientific mosaic"?
- What are the indicators of a *scientific community*, i.e. how does one determine that a community in question is scientific?
- When studying a certain community in a certain time period, how much of their mosaic should one *reconstruct*? If only some parts of it, then how does one decide which parts to focus on and which parts to omit?

Theoretical Questions

- What is the place of *problems* (issues, questions) in the scientific mosaic? What role do problems play in scientific change?
- What is the status of *normative* propositions (e.g. ethics, methodology, etc.)? Are they part of the scientific mosaic? If so, what is the mechanism of their transformations? Are they subject to the laws of scientific change?
- Is there an explanation for changes in *disciplinary boundaries*? Is there any separate mechanism for these changes, or are they by-products of more fundamental processes of theory and method change, governed by the laws of scientific change? Do disciplinary boundaries play any considerable role in the process of scientific change?
- What happens when the community realises that a theory was accepted by *mistake* (e.g. it appears to have provided confirmed novel predictions, but then it turns out that it hasn't)? Is there a change-revoking procedure in science? How are the erroneous acceptances handled?

© Springer International Publishing Switzerland 2015
H. Barseghyan, *The Laws of Scientific Change*, DOI 10.1007/978-3-319-17596-6

- Can the *sociocultural factors* play any role in *method* employment? If so, how exactly can they affect the process?
- Is *the second law* really a tautology, or does it have some empirical content after all?
- Is *the zeroth law* a tautology? If not, what kind of historical events can possibly violate *the zeroth law*?

Historical Questions

- What were the *criteria of compatibility* at different time periods and different mosaics? How are they to be explicated?
- How did *disciplinary boundaries* change through time?
- How and why was *astrology* exiled from the mosaic? What was the cause of the exile?
- Was *theology* exiled from the mosaic, or was it an instance of mosaic split? If the latter, what is the history of that mosaic (e.g. when was heliocentrism actually accepted into the Catholic mosaic)?
- When and how was the Aristotelian-Cartesian requirement of *intuitive truth* abandoned?
- Ethical beliefs have clearly affected theory construction and pursuit. But have there been any cases when *ethical* beliefs affected theory acceptance/rejection? If so, were those moral considerations part of the method of the time?
- When and by what theory was the *humorist (Galenic) medicine* replaced in the mosaic? Was this process different in different mosaics?

Philosophical Questions

- What is the *ontological* status of the laws of scientific change? Under what social, political, and/or economic conditions do the patterns of scientific change emerge and hold?
- Are the laws of scientific change *reducible* to some sociological or psychological laws?
- Can the TSC be used to solve the problem of *demarcation* between science and non-science?
- Can the TSC be used to solve the problem of *scientific progress*?

Legend

Logical Relations:
Note: For convenience, the respective truth tables are also included:

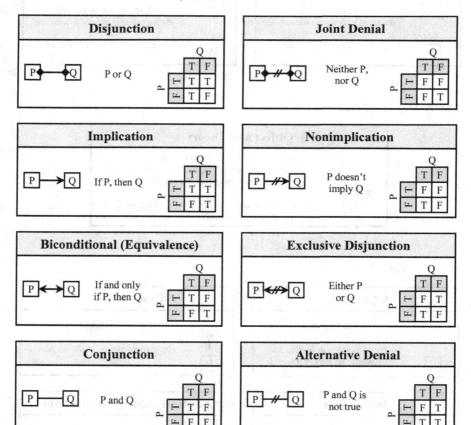

© Springer International Publishing Switzerland 2015
H. Barseghyan, *The Laws of Scientific Change*, DOI 10.1007/978-3-319-17596-6

Process and Governing Laws:

Temporal Succession (Process)
S_1 S_2 Process S: state S_2 succeeds S_1 *E.g. one state of the mosaic follows another*

Governing (Objective) Laws
Process S is governed by law L S_1 S_2 *E.g. physical processes are governed by respective laws*

Theory and Method:

Theory
Folded: A set of propositions that attempts to describe something.   *Unfolded*:   T   *E.g. general relativity*

Method
Folded: A set of requirements for employment in theory assessment.   *Unfolded*:   M   *E.g. the blind trial method*

Theory-Object relation:

Object and Theory
Theory T accounts for process S. S is an object of theory T. In this relation, T is at the theory-level, while process S is at the object-level.   *E.g. changes in the mosaic are the object of theory of scientific change.*

UML symbols:

Aggregation
A   B B can be part of A *E.g. a theory can be (but not necessarily is) part of a mosaic*

Composition
A   B B is part of (belongs to) A *E.g. a requirement is always part of some method*

Inheritance
A   B B is a subtype of A *E.g. substantive method is a subtype of method*

Implementation (Realization)
A   B B is an implementation of A *E.g. a concrete method implements (specifies) an abstract method*

Bibliography

Abímbólá, K. (2006). Rationality and methodological change: Dudley Shapere's conception of scientific development. *Principia, 10*, 39–65.

Agassi, J. (1972). *Faraday as a natural philosopher*. Chicago: Chicago University Press.

Aiton, E. J. (1958). The vortex theory of the planetary motions – III. *Annals of Science, 14*, 157–172.

Aiton, E. J. (1989). The Cartesian vortex theory. In Taton and Wilson (eds.) (1989), pp. 207–221.

Albert, D. Z. (1992). *Quantum mechanics and experience*. Cambridge, MA: Harvard University Press.

Allchin, D. (2003). Lawson's Shoehorn, or should the philosophy of science be rated 'X'? *Science and Education, 12*, 315–329.

Amorim Neto, O., & Cox, G. W. (1997). Electoral institutions, cleavage structures, and the number of parties. *American Journal of Political Science, 41*(1), 149–174.

Arabatzis, T. (2006). On the inextricability of the context of discovery and the context of justification. In Schickore and Steinle (eds.) (2006), pp. 215–230.

Ariew, R. (1992). Theory of comets at Paris during the seventeenth century. *Journal of the History of Ideas, 53*, 355–372.

Armitage, A. (1950). Rene Descartes (1596–1650) and the early royal society. *Notes and Records of the Royal Society of London, 8*, 1–19.

Barberousse, A. (2008). From one version to the other: Intra-theoretical change. In L. Soler, Sankey and Hoyningen-Huene (eds.) (2008), pp. 87–101.

Barnes, B., & Bloor, D. (1982). Relativism, rationalism and the sociology of knowledge. In Hollis and Lukes (eds.) (1982), pp. 21–47.

Barseghyan, H. (2009). The regularity-oriented approach to the ontological problem of reduction. Unpublished Research Paper.

Bennett, J. A. (1989). Magnetical philosophy and astronomy from Wilkins to Hooke. In Taton and Wilson (eds.) (1989), pp. 222–230.

Bernard, G. W. (2005). *The king's reformation: Henry VIII and the remaking of the English church*. New Haven: Yale University Press.

Biagioli, M. (1996). From relativism to contingentism. In Galison and Stump (eds.) (1996), pp. 189–206.

Blachowicz, J. (2009). How science textbooks treat scientific method: A philosopher's perspective. *The British Journal for the Philosophy of Science, 60*, 303–334.

Boas, M. (1952). The establishment of the mechanical philosophy. *Osiris, 10*, 412–541.

Bourg, J. (2001). The rhetoric of modal equivocacy in Cartesian transubstantiation. *Journal of the History of Ideas, 62*, 121–140.

Brigandt, I., & Love, A. (2008). Reductionism in biology. In Zalta (ed.) (2013). http://plato.stanford.edu/archives/win2013/entries/reduction-biology/

© Springer International Publishing Switzerland 2015

H. Barseghyan, *The Laws of Scientific Change*, DOI 10.1007/978-3-319-17596-6

Brock, W. A., & Durlauf, S. N. (1999). A formal model of theory choice in science. *Economic Theory, 14*, 113–130.

Brockliss, L. (1981). Aristotle, Descartes and the new science: Natural philosophy at the University of Paris, 1600–1740. *Annals of Science, 38*, 33–69.

Brockliss, L. (2003). Science, the universities, and other public spaces: Teaching science in Europe and the Americas. In Porter (ed.) (2003), pp. 44–86.

Brockliss, L. (2006). The moment of no return: The University of Paris and the death of Aristotelianism. *Science and Education, 15*, 259–278.

Brooke, J. H. (1991). *Science and religion. Some historical perspectives*. Cambridge: Cambridge University Press.

Brown, J. R. (2001). *Who rules in science?* Cambridge, MA: Harvard University Press.

Brush, S. G. (1994). Dynamics of theory change: The role of predictions. *Proceedings of the Biennial Meeting of the PSA, 2*, 133–145.

Buchwald, J. Z., & Franklin, A. (2005). Introduction: Beyond disunity and historicism. In Buchwald and Franklin (eds.) (2005), pp. 1–16.

Buchwald, J. Z., & Franklin, A. (Eds.). (2005). *Wrong for the right reasons*. Dordrecht: Springer.

Bueno, O. (2008). Structural realism, scientific change, and partial structures. *Studia Logica, 89*, 213–235.

Bueno, O., da Costa, N. C. A., & French, S. (1998). The logic of pragmatic truth. *Journal of Philosophical Logic, 27*, 603–620.

Bunge, M. (1962). *Intuition and science*. Englewood Cliffs, NJ: Prentice-Hall.

Bunge, M. (1973). *Philosophy of physics*. Boston: D. Reidel Publishing Company.

Bunge, M. (1974). *Treatise on basic philosophy. Volume 1. Semantics I: Sense and reference*. Dordrecht: D. Reidel Publishing Company.

Bunge, M. (1983). *Treatise on basic philosophy. Volume 6. Epistemology and methodology II: Understanding the world*. Boston: D. Reidel Publishing Company.

Bunge, M. (1998). *Social science under debate. A philosophical perspective*. Toronto: University of Toronto Press.

Burgess, J. P. (2009). *Philosophical logic*. Princeton: Princeton University Press.

Butterfield, H. (1931). *The Whig interpretation of history*. London: G. Bell and Sons, 1963.

Campion, N. (2009). *A history of western astrology. Volume II: The medieval and modern worlds*. London: Continuum.

Cartwright, N. (1999). *The dappled world. A study of the boundaries of science*. Cambridge, UK: Cambridge University Press.

Chakravartty, A. (2007). *A metaphysics for scientific realism*. Cambridge: Cambridge University Press.

Chang, H. (2004). *Inventing temperature. Measurement and scientific progress*. Oxford: Oxford University Press.

Clarke, D. S. (1994). Does acceptance entail belief? *American Philosophical Quarterly, 31*, 145–155.

Cohen, I. B. (1985). *The birth of a new physics*. New York: W. W. Norton & Company.

Cohen, I. B., & Smith, G. E. (Eds.). (2002). *The Cambridge companion to Newton*. Cambridge, UK: Cambridge University Press.

Cummings, B. (Ed.). (2011). *The book of common prayer. The texts of 1549, 1559, 1662*. Oxford: Oxford University Press.

Cushing, J. T., Delaney, C. F., & Gutting, G. M. (1984). *Science and reality. Recent work in the philosophy of science. Essays in honor of Ernan McMullin*. Notre Dame: University of Notre Dame Press.

d'Alembert, J. L. R. (1751). *Preliminary discourse to the encyclopedia of Diderot*. Chicago: The University of Chicago press, 1995.

d'Espagnat, B. (2008). Is science cumulative? A physicist viewpoint. In Soler, Sankey and Hoyningen-Huene (eds.) (2008), pp. 145–151.

Dales, R. C. (1973). *The scientific achievement of the middle ages*. Philadelphia: University of Pennsylvania Press, 1989.

Dear, P. (2005). What is the history of science the history of? Early modern roots of the ideology of modern science. *Isis, 96*, 390–406.

Diderot, D. (Ed.) (1751–1780). *Encyclopédie ou Dictionnaire raisonné des sciences, des arts et des métiers*. Stuttgart: F. Frommann, 1966–1967.

Donovan, A., Laudan, L., & Laudan, R. (Eds.). (1992). *Scrutinizing science. Empirical studies of scientific change*. Baltimore: The Johns Hopkins University Press.

Doppelt, G. (1990). The naturalist conception of methodological standards. *Philosophy of Science, 57*, 1–19.

Duverger, M. (1954). *Political parties*. New York: Wiley.

Easton, P. (2005). Desgabets's indefectibility thesis – A step too far? In Schmaltz (ed.) (2005), pp. 27–41.

Einstein, A. (1934). On the method of theoretical physics. *Philosophy of Science, 1*, 163–169.

Einstein, A. (1953). Foreword. In Galileo (1632), pp. vi–xx.

Elster, J. (2007). *Explaining social behavior: More nuts and bolts for the social sciences*. Cambridge, UK: Cambridge University Press.

Evans, D. S. (1958). Dashing and dutiful. *Science, 127*(3304), 935–948. New Series.

Evans, D. S. (1967). Historical notes on astronomy in South Africa. *Vistas in Astronomy, 9*, 265–282.

Evans, D. S. (1992). *Lacaille: Astronomer, traveler. With a new translation of his journal*. Tucson: Pachart Publishing House.

Fara, P. (2003). Marginalized practices. In Porter (ed.) (2003), pp. 285–507.

Fatigati, M. (2014). A method for reconstructing the medieval Arabic scientific mosaic. Unpublished.

Feyerabend, P. S. (1970). Consolations for the specialist. In Lakatos and Musgrave (eds.) (1970), pp. 197–230.

Feyerabend, P. S. (1975). *Against method*. London: Verso, 2002.

Feyerabend, P. S. (1981). *Realism, rationalism and scientific method: Philosophical papers volume 1*. Cambridge, UK: Cambridge University Press.

Fieser, J., & Dowden, B. (Eds.). (2014). The internet encyclopedia of philosophy. http://www.iep. utm.edu/

Fine, A. (1982). Antinomies of entanglement: The puzzling case of the tangled statistics. *The Journal of Philosophy, 79*, 733–747.

Finocchiaro, M. (1992). Galileo's Copernicanism and the acceptability of guiding assumptions. In Donovan, Laudan and Laudan (eds.) (1992), pp. 49–67.

Frängsmyr, T. (1974). Swedish science in the eighteenth century. *History of Science, 12*, 29–42.

Freedman, K. L. (2009). Normative naturalism and epistemic relativism. *International Studies in the Philosophy of Science, 20*, 309–322.

Friedman, M. (2001). *Dynamics of reason*. Stanford: CSLI Publications.

Galileo, G. (1632). *Dialogue concerning the two chief world systems* (trans: Drake, S.). Berkeley: University of California Press, 1967.

Galison, P. (2008). Ten problems in history and philosophy of science. *Isis, 99*, 111–124.

Galison, P., & Stump, D. J. (Eds.). (1996). *The disunity of science: Boundaries, contexts, and power*. Stanford: Stanford University Press.

Garber, D. (1986). Learning from the past: Reflections on the role of history in the philosophy of science. *Synthese, 67*, 91–114.

Garber, D. (1992). *Descartes' metaphysical physics*. Chicago: The University of Chicago Press.

Gardner, M. R. (1982). Predicting novel facts. *British Journal for the Philosophy of Science, 33*, 1–15.

Gascoigne, J. (1989). *Cambridge in the age of the enlightenment. Science, religion and politics from the restoration to the French Revolution*. Cambridge: Cambridge University Press.

Ghirardi, G. (2005). *Sneaking a look at God's cards. Unraveling the mysteries of quantum mechanics*. Princeton: Princeton University Press.

Giere, R. N. (1973). Review: History and philosophy of science: Intimate relationship or marriage of convenience? *The British Journal for the Philosophy of Science, 24*, 282–297.

Giere, R. N. (1984). Toward a unified theory of science. In Cushing et al. (1984), pp. 5–31.

Gilbert, M. (1987). Modelling collective belief. *Synthese, 73*, 185–204.

Glennan, S. (2010). Ephemeral mechanisms and historical explanation. *Erkenntnis, 72*, 251–266.

Godfrey-Smith, P. (2003). *Theory and reality.* Chicago: The University of Chicago Press.

Gorton, W. A. (2014). The philosophy of social science. In Fieser and Dowden (2014). http://www. iep.utm.edu/soc-sci/#SH2a

Grant, E. (2004). *Science and religion. 400 BC – AD 1550.* Baltimore: The Johns Hopkins University Press.

Greenberg, J. L. (1987). The measurement of the earth. Essay review. *Annals of Science, 44*, 289–295.

Greenberg, J. L. (1995). *The problem of the earth's shape from Newton to Clairaut: The rise of mathematical science in eighteenth-century Paris and the fall of "Normal" science.* Cambridge: Cambridge University Press.

Guerlac, H. (1981). *Newton on the continent.* Ithaca: Cornell University Press.

Hacking, I. (1996). The disunities of science. In Galison and Stump (eds.) (1996), pp. 37–74.

Hacking, I. (1999). *The social construction of what?* Cambridge, MA: Harvard University Press.

Hansson, S. O. (2008). Science and pseudo-science. In Zalta (ed.) (2013). http://plato.stanford. edu/archives/win2013/entries/pseudo-science/

Hansson, S. O. (2013). Defining science and pseudoscience. In Pigliucci & Boudry (Eds.), *Philosophy of pseudoscience* (pp. 61–77).

Hempel, C. G. (1942). The function of general laws in history. *The Journal of Philosophy, 39*, 35–48.

Hitchcock, C. (Ed.). (2004). *Contemporary debates in philosophy of science.* Malden: Blackwell Publishing.

Hollis, M., & Lukes, S. (Eds.). (1982). *Rationality and relativism.* Cambridge, MA: The MIT Press.

Horwich, P. (1991). On the nature and norms of theoretical commitment. *Philosophy of Science, 58*, 1–14.

Howson, C., & Urbach, P. (2006). *Scientific reasoning. The Bayesian approach.* La Salle: Open Court.

Hoyningen-Huene, P. (2006). Context of discovery versus context of justification and Thomas Kuhn. In Schickore and Steinle (eds.) (2006), pp. 119–131.

Hudson, R. G. (2007). What's really at issue with novel predictions? *Synthese, 155*, 1–20.

Hull, D. L. (1979). In defense of presentism. *History and Theory, 18*, 1–15.

Hume, D. (1739/40). *A treatise of human nature.* Oxford: Oxford University Press, 2007.

Huntington, S. P. (1968). *Political order in changing societies.* New Haven: Yale University Press.

Iliffe, R. (2003). Philosophy of science. In Porter (ed.) (2003), pp. 267–284.

Jaki, S. L. (1978). Johann Georg von Soldner and the gravitational bending of light, with an English translation of his essay on it published in 1801. *Foundations of Physics, 8*, 927–950.

John Paul II. (2003). *Encyclical Letter. Ecclesia De Eucharistia.* http://w2.vatican.va/content/john-paul-ii/en/encyclicals/documents/hf_jp-ii_enc_20030417_eccl-de-euch.html.

Jorink, E., & Maas, A. (Eds.). (2012). *Newton & the Netherlands: How Isaac Newton was fashioned in the Dutch republic.* Amsterdam: Leiden University Press.

Kaku, M. (2011). Breaking the speed of light and contemplating the demise of relativity. http://bigthink.com/dr-kakus-universe/breaking-the-speed-of-light-and-contemplating-the-demise-of-relativity

Kantorowicz, E. (1957). *The king's two bodies: A study in mediaeval political theology.* Princeton: Princeton University Press, 1997.

Kelly, K. T. (2000). Naturalism logicized. In Nola and Sankey (eds.) (2000), pp. 177–210.

Kieseppä, I. A. (2000). Rationalism, naturalism, and methodological principles. *Erkenntnis, 53*, 337–352.

Kincaid, H. (1990). Defending laws in the social sciences. *Philosophy of the Social Sciences, 20*, 56–83.

Kincaid, H. (2004). There are laws in the social sciences. In Hitchcock (ed.) (2004), pp. 168–185.

King, G., Keohane, R. O., & Verba, S. (1994). *Designing social inquiry: Scientific inference in qualitative research*. Princeton: Princeton University Press.

Kitcher, P., & Salmon, W. C. (1989). *Scientific explanation*. Minneapolis: University of Minnesota Press.

Knowles, J. (2002). What's really wrong with Laudan's normative naturalism. *International Studies in the Philosophy of Science, 16*, 171–186.

Koertge, N. (1980). Analysis as a method of discovery during the scientific revolution. In Nickles (ed.) (1980), pp. 139–157.

Kolman, J., Meng, P., & Scott, G. (Eds.). (1998). *Good clinical practice: Standard operating procedures for clinical researchers*. Chichester: Wiley.

Koyré, A. (1957). *From the closed world to the infinite universe*. Baltimore: The Johns Hopkins Press.

Kragh, H. (1999). *Quantum generations: A history of physics in the twentieth century*. Princeton: Princeton University Press.

Krementsov, N. (1997). *Stalinist science*. Princeton: Princeton University Press.

Kuhn, H. (2005). Aristotelianism in renaissance. In Zalta (ed.) (2013). http://plato.stanford.edu/archives/win2013/entries/aristotelianism-renaissance/

Kuhn, T. S. (1957). *The Copernican revolution. Planetary astronomy in the development of western thought*. Cambridge, MA: Harvard University Press, 1985.

Kuhn, T. S. (1962/70) *The structure of scientific revolutions*. Chicago: The University of Chicago Press, 1996.

Kuhn, T. S. (1970a). Reflections on my critics. In Lakatos and Musgrave (eds.), (1970) pp. 231–278.

Kuhn, T. S. (1970b). Notes on Lakatos. *Proceedings of the Biennial Meeting of the PSA, 1970*, 137–146.

Kuhn, T. S. (1977). *The essential tension. Selected studies in scientific tradition and change*. Chicago: The University of Chicago Press.

Kuhn, T. S. (2000). *The road since structure*. Chicago: The University of Chicago Press.

Lacey, H. (2004). *Is science value free? Values and scientific understanding*. London: Routledge.

Lafuente, A., & Delgado, A. (1984). *La Geometrizacion de la Tierra: Observaciones y Resultados de la Expedición Geodésica Hispano-Francesa al Virreinato del Perú (1735–1744)*. Madrid: Instituto Arnau de Villanova.

Lakatos, I. (1968). Changes in the problem of inductive logic. In Lakatos (1978b), pp. 128–200.

Lakatos, I. (1970). Falsification and the methodology of scientific research programmes. In Lakatos (1978a), pp. 8–101.

Lakatos, I. (1971). History of science and its rational reconstructions. In Lakatos (1978a), pp. 102–138.

Lakatos, I. (1978a). *Philosophical papers, volume I*. Cambridge: Cambridge University Press.

Lakatos, I. (1978b). *Philosophical papers, volume II*. Cambridge: Cambridge University Press.

Lakatos, I., & Musgrave, A. (Eds.). (1970). *Criticism and the growth of knowledge*. Cambridge: Cambridge University Press.

Lakatos, I., & Zahar, E. (1976). Why did Copernicus's research programme supersede Ptolemy's? In *Philosophical papers, volume 1* (pp. 168–192). Cambridge: Cambridge University Press, 1978.

Langley, P., Simon, H. A., Bradshaw, G. L., & Zytkow, J. M. (1987). *Scientific discovery: Computational explorations of the creative processes*. Cambridge, MA: The MIT Press.

Latour, B., & Woolgar, S. (1986). *Laboratory life: The construction of scientific facts*. Princeton: Princeton University Press.

Laudan, L. (1968). Theories of scientific method from Plato to Mach. A bibliographical review. *History of Science, 7*, 1–63.

Laudan, L. (1977). *Progress and its problems. Toward a theory of scientific growth*. Berkeley: University of California Press.

Laudan, L. (1980). Why was the logic of discovery abandoned? In Nickles (ed.), (1980) pp. 173–183.

Laudan, L. (1981). *Science and hypothesis. Historical essays on scientific methodology*. Dordrecht: D. Reidel Publishing Company.

Laudan, L. (1984). *Science and values*. Berkeley: University of California Press.

Laudan, L. (1989). If it ain't broke don't fix it. *The British Journal for the Philosophy of Science, 40*, 369–375.

Laudan, L. (1990). The history of science and the philosophy of science. In Olby et al. (eds.) (1990), pp. 47–59.

Laudan, L. (1996). *Beyond positivism and relativism. Theory, method and evidence*. Boulder: Westview Press.

Laudan, R., & Laudan, L. (1989). Dominance and the disunity of method: Solving the problems of innovation and consensus. *Philosophy of Science, 56*, 221–237.

Laudan, L., & Leplin, J. (1991). Empirical equivalence and under determination. *Journal of Philosophy, 88*, 449–472.

Laudan, L., et al. (1986). Scientific change: Philosophical models and historical research. *Synthese, 69*, 141–223.

Lawson, A. E. (2002). What does Galileo's discovery of Jupiter's moons tell us about the process of scientific discovery? *Science and Education, 11*, 1–24.

Leplin, J. (1990). Renormalizing naturalism. *Philosophy of Science, 57*, 20–33.

Leplin, J. (1997). *A novel defence of scientific realism*. New York: Oxford University Press.

Lindberg, D. (2008). *The beginnings of western science*. Chicago: The University of Chicago Press.

Little, D. (1993). On the scope and limits of generalizations in the social sciences. *Synthese, 97*(2), 183–207.

Lovejoy, A. O. (1936). *The great chain of being*. Cambridge, MA: Harvard University Press, 1964.

Lugg, A. (1984). Review: Changing fortunes of the method of hypothesis. *Erkenntnis, 21*, 433–438.

Lycan, W. G. (1988). *Judgement and justification*. Cambridge: Cambridge University Press.

Maclear, T. (1866). *Verification and extension of Lacaille's arc of the meridian at the Cape of Good Hope*. London: S.I.

Maienschein, J., & Smith, G. (2008). What difference does history of science make, anyway? *Isis, 99*, 318–321.

Mayr, E. (1990). When is historiography whiggish? *Journal of the History of Ideas, 51*, 301–309.

McClaughlin, T. (1979). Censorship and defenders of the Cartesian faith in mid-seventeenth century France. *Journal of the History of Ideas, 40*, 563–581.

McIntyre, L. C. (1993). Complexity and social scientific laws. *Synthese, 97*(2), 209–227.

McIntyre, L. C. (1998). *Laws and explanation in the social sciences*. Boulder: Westview Press.

McMullin, E. (1988). The shaping of scientific rationality: Construction and constraint. In McMullin (ed.) (1988), pp. 1–47.

McMullin, E. (Ed.). (1988). *Construction and constraint. The shaping of scientific rationality*. Notre Dame: University of Notre Dame Press.

McMullin, E. (Ed.). (1992). *The social dimensions of science*. Notre Dame: University of Notre Dame Press.

Medawar, P. B. (1979). *Advice to a young scientist*. New York: Basic Books.

Meheus, J. (2003). Inconsistencies and the dynamics of science. *Logic and Logical Philosophy, 11*, 129–148.

Motterlini, M. (Ed.). (1999). *For and against method. Imre Lakatos and Paul Feyerabend*. Chicago: The University of Chicago Press.

Motterlini, M. (2002). Reconstructing Lakatos: A reassessment of Lakatos' epistemological project in the light of the Lakatos archive. *Studies in History and Philosophy of Science, 33*, 487–509.

Musgrave, A. (1974). Logical versus historical theories of confirmation. *British Journal for the Philosophy of Science, 25*, 1–23.

Nadler, S. M. (1988). Arnauld, Descartes, and transubstantiation: Reconciling Cartesian metaphysics and real presence. *Journal of the History of Ideas, 49*, 229–246.

Neurath, O. (1973). *Empiricism and sociology*. Dordrecht: D. Reidel Publishing Company.

Newton, I. (1687). *The principia. Mathematical principles of natural philosophy*. Berkeley: University of California Press, 1999.

Newton-Smith, W. H. (1981). *The rationality of science*. Boston: Routledge & Kegan Paul.

Nickles, T. (1980). Introductory essay: Scientific discovery and the future of philosophy of science. In Nickles (ed.) (1980), pp. 1–59.

Nickles, T. (Ed.). (1980). *Scientific discovery, logic and rationality*. Dordrecht: D Reidel Publishing Company.

Nickles, T. (1986). Remarks on the use of history as evidence. *Synthese, 69*, 253–266.

Nickles, T. (1989). Review: Historicism and scientific practice I. *Isis, 80*, 665–669.

Nickles, T. (1992). Good science as bad history: From order of knowing to order of being. In McMullin (ed.) (1992), pp. 85–129.

Nickles, T. (1995). Philosophy of science and history of science. *Osiris, 10*, 138–163 (Constructing knowledge in the history of science).

Nickles, T. (2006). Heuristic appraisal: Context of discovery or justification? In Schickore and Steinle (eds.) (2006), pp. 159–182.

Nola, R., & Sankey, H. (2000). A selective survey of theories of scientific method. In Nola and Sankey (eds.) (2000), pp. 1–65.

Nola, R., & Sankey, H. (Eds.). (2000). *After Popper, Kuhn and Feyerabend. Recent issues in theories of scientific method*. Dordrecht: Kluwer.

Nola, R., & Sankey, H. (2007). *Theories of scientific method*. Montreal: McGill-Queens's University Press.

Olby, R. C., et al. (Eds.). (1990). *Companion to the history of modern science*. London: Routledge.

Olson, R. G. (2004). *Science and religion, 1450–1900: From Copernicus to Darwin*. Baltimore: The Johns Hopkins University Press.

Pagden, A. (1988). The reception of the 'New Philosophy' in eighteenth-century Spain. *Journal of the Warburg and Courtauld Institutes, 51*, 126–140.

Peckhaus, V. (2006). Psychologism and the distinction between discovery and justification. In Schickore and Steinle (eds.) (2006), pp. 99–116.

Penrose, R. (1999). The central programme of twistor theory. *Chaos, Solitons & Fractals, 10*, 581–611.

Penrose, R. (2004). *The road to reality: A complete guide to the physical universe*. London: Vintage Books.

Pickering, A. (1984). *Constructing quarks. A sociological history of particle physics*. Chicago: The University of Chicago Press.

Pigliucci, M., & Boudry, M. (Eds.). (2013). *Philosophy of pseudoscience: Reconsidering the demarcation problem*. Chicago: University of Chicago Press.

Pinnick, C., & Gale, G. (2000). Philosophy of science and history of science: A troubling interaction. *Journal for General Philosophy of Science, 31*, 109–125.

Popper, K. R. (1934/59). *The logic of scientific discovery*. London: Routledge, 2006.

Popper, K. R. (1963). *Conjectures and refutations*. London: Routledge, 2007.

Porter, T. M. (1991). The uses of humanistic history. *Philosophy of the Social Sciences, 21*, 214–222.

Porter, R. (Ed.). (2003). *The Cambridge history of science. Volume 4: Eighteenth-century science*. Cambridge: Cambridge University Press.

Preston, J. (1994). Review: Scrutinizing science: Empirical studies of scientific change by Arthur Donovan; Larry Laudan; Rachel Laudan. *The British Journal for the Philosophy of Science, 45*, 1063–1065.

Price, M. C. (1995). The Everett FAQ. http://www.hedweb.com/manworld.htm

Pumfrey, S. (1989). Magnetical philosophy and astronomy, 1600–1650. In Taton and Wilson (eds.) (1989), pp. 45–53.

Quine, W. V. O., & Ullian, J. S. (1978). *The web of belief*. New York: McGraw-Hill.

Richardson, A. W. (1992). Philosophy of science and its rational reconstructions: Remarks on the VPI program for testing philosophies of science. *Proceedings of the Biennial Meeting of the PSA, 1*, 36–46.

Rorty, R. (1988). Is natural science a natural kind? In McMullin (ed.) (1988), pp. 49–74.

Rosenberg, A. (1990). Normative naturalism and the role of philosophy. *Philosophy of Science, 57*, 34–43.

Rovelli, C. (2000). *Notes for a brief history of quantum gravity.* Presented at the 9th Marcel Grossmann Meeting in Roma, July 2000. http://arxiv.org/pdf/gr-qc/0006061v3.pdf

Russell, B. (1959). *The ABC of relativity.* New York: A Mentor Book, 1985.

Sagan, C. (presenter) (1980). *Cosmos: A personal voyage. Episode 12: Encyclopaedia galactica.* Dir. Adrian Malone. PBS.

Salmon, M. H. (1989). Explanation in the social sciences. In Kitcher and Salmon (eds.) (1989), pp. 384–409.

Schickore, J., & Steinle, F. (Eds.). (2006). *Revisiting discovery and justification.* Dordrecht: Springer.

Schlesinger, G. N. (1987). Accommodation and prediction. *Australasian Journal of Philosophy, 65*, 33–42.

Schmaltz, T. M. (1996). *Malebranche's theory of soul. A Cartesian interpretation.* New York: Oxford University Press.

Schmaltz, T. M. (2005). French Cartesianism in context: The Paris formulary and Regis's usage. In Schmaltz (ed.) (2005), pp. 80–95.

Schmaltz, T. M. (Ed.). (2005). *Receptions of Descartes. Cartesianism and anti-Cartesianism in early modern Europe.* London: Routledge.

Schmitt, C. B. (1973). Towards a reassessment of renaissance Aristotelianism. *History of Science, 11*, 159–193.

Schofield, C. (1989). The Tychonic and semi-Tychonic world systems. In Taton and Wilson (eds.) (1989), pp. 33–44.

Shapere, D. (1980). The character of scientific change. In Nickles (ed.) (1980), pp. 61–116.

Shapin, S. (1996). *The scientific revolution.* Chicago: The University of Chicago Press.

Siegel, H. (1990). Laudan's normative naturalism. *Studies in the History and Philosophy of Science, 21*, 295–313.

Simon, H. A. (1992). Scientific discovery as problem solving: Reply to critics. *International Studies in the Philosophy of Science, 6*, 69–88.

Smith, J. M. (1988). Inconsistency and scientific reasoning. *Studies in History and Philosophy of Science, 19*, 429–445.

Smith, G. E. (1989). The methodology of the principia. In Cohen and Smith (eds.) (2002), pp. 138–173.

Soler, L., Sankey, H., & Hoyningen-Huene, P. (Eds.). (2008). *Rethinking scientific change and theory comparison: Stabilities, ruptures, incommensurabilities?* Dordrecht: Springer.

Solomon, M. (1992). Scientific rationality and human reasoning. *Philosophy of Science, 59*, 439–455.

Solomon, M. (1994). Social empiricism. *Noûs, 28*, 325–343.

Stanford, K. (2013). Underdetermination of scientific theory. In Zalta (ed.). (2013) http://plato.stanford.edu/archives/win2013/entries/scientific-underdetermination/

Steinle, F. (2006). Concept formation and the limits of justification: "Discovering" the two electricities. In Schickore and Steinle (eds.) (2006), pp. 183–195.

Stocking, G. W., Jr. (1965). On the limits of 'Presentism' and 'Historicism' in the historiography of the behavioral sciences. *Journal of the History of the Behavioral Sciences, 1*, 211–218.

Stoljar, D. (2009). Physicalism. In Zalta (ed.) (2013). http://plato.stanford.edu/archives/win2013/entries/physicalism/

Sturm, T., & Gigerenzer, G. (2006). How can we use the distinction between discovery and justification? On the weaknesses of the strong programme in the sociology of science. In Schickore and Steinle (eds.) (2006), pp. 133–158.

Suppe, F. (1989). *The semantic conception of theories and scientific realism.* Chicago: University of Illinois Press.

Taton, R., & Wilson, C. (Eds.). (1989). *The general history of astronomy. Volume 2: Planetary astronomy from the renaissance to the rise of astrophysics. Part A: Tycho Brahe to Newton.* Cambridge: Cambridge University Press, 2003.

Tegmark, M. (1998). The interpretation of quantum mechanics: Many worlds or many words? *Fortschritte der Physik, 46,* 855–862.

Terrall, M. (1992). Representing the earth's shape: The polemics surrounding Maupertuis's expedition to Lapland. *Isis, 83,* 218–237.

Terrall, M. (2002). *The man who flattened the earth. Maupertuis and the sciences in the enlightenment.* Chicago: The University of Chicago Press, 2006.

Thoren, V. (1989). Tycho Brahe. In Taton and Wilson (eds.), (1989) pp. 3–21.

Thornton, S. (2009). Karl Popper. In Zalta (ed.) (2013). http://plato.stanford.edu/archives/win2013/entries/popper/

Truesdell, C. (1960). A program toward rediscovering the rational mechanics of the age of reason. *Archive for the History of Exact Sciences, 1,* 3–36.

Turner, D. M. (1927). *History of science teaching in England.* London: Chapman & Hall.

Uebel, T. (2011). Vienna circle. In Zalta (ed.) (2013). http://plato.stanford.edu/archives/win2013/entries/vienna-circle/

Van Fraassen, B. C. (1980). *The scientific image.* Oxford: Clarendon Press, 1995.

Van Helden, A. (1989). Galileo, telescopic astronomy, and the Copernican system. In Taton and Wilson (eds.) (1989), pp. 81–105.

Van Helden, A. (1994). Telescopes and authority from Galileo to Cassini. *Osiris, 9,* 8–29.

Vartanian, A. (1953). *Diderot and Descartes. A study of scientific naturalism in the enlightenment.* Princeton: Princeton University Press.

Voltaire, F.-M. A. (1733). *Letters on England.* London: Penguin Books. 2005.

Weinberg, S. (1992). *Dreams of a final theory.* New York: Vintage Books.

Weinberg, S. (2003). *Facing up: Science and its cultural adversaries.* Cambridge, MA: Harvard University Press.

Westfall, R. S. (1971). *The construction of modern science: Mechanisms and mechanics.* Cambridge: Cambridge University Press.

Westman, R. S. (1975). The Melanchthon circle, Rheticus, and the Wittenberg interpretation of the Copernican theory. *Isis, 66,* 165–193.

Whewell, W. (1840). *The philosophy of the inductive sciences, founded upon their history* (Vol. 1). New York: Johnson Reprint Corp. 1967.

Whewell, W. (1860). *On the philosophy of discovery.* Whitefish: Kessinger Publishing. 2009.

Whitt, L. A. (1990). Theory pursuit: Between discovery and acceptance. *Proceedings of the Biennial Meeting of the PSA, 1,* 467–483.

Will, C. M. (1988). Henry Cavendish, Johann von Soldner, and the deflection of light. *American Journal of Physics, 56,* 413–415.

Williams, P. L. (1975). Should philosophers be allowed to write history? *The British Journal for the Philosophy of Science, 26,* 241–253.

Wilson, C. (1989a). Predictive astronomy in the century after Kepler. In Taton and Wilson (eds.) (1989), pp. 161–206.

Wilson, C. (1989b). The Newtonian achievement in astronomy. In Taton and Wilson (eds.) (1989), pp. 233–274.

Wilson, A., & Ashplant, T. G. (1988a). Whig history and present-centred history. *The Historical Journal, 31,* 1–16.

Wilson, A., & Ashplant, T. G. (1988b). Present-centred history and the problem of historical knowledge. *The Historical Journal, 31,* 253–274.

Worrall, J. (1988). Review: The value of a fixed methodology. *The British Journal for the Philosophy of Science, 39,* 263–275.

Worrall, J. (1989). Fix it and be damned: A reply to Laudan. *The British Journal for the Philosophy of Science, 40,* 376–388.

Worrall, J., & Scerri, E. R. (2001). Prediction and the periodic table. *Studies in History and Philosophy of Science, 32,* 407–452.

Wray, B. K. (2001). Collective belief and acceptance. *Synthese, 129,* 319–333.

Wykstra, S. J. (1980). Toward a historical meta-method for assessing normative methodologies: Rationability, serendipity, and the Robinson Crusoe Fallacy. *Proceedings of the Biennial Meeting of the PSA, 1,* 211–222.

Yeo, R. (2003). Classifying the sciences. In Porter (ed.), (2003), pp. 241–266.

Zahar, E. (1973). Why did Einstein's programme supersede Lorentz's? *The British Journal for the Philosophy of Science, 24,* 95–123.

Zahar, E. (1982). Review: Feyerabend on observation and empirical content. *The British Journal for the Philosophy of Science, 33,* 397–409.

Zahar, E. (1989). *Einstein's revolution: A study in heuristic.* La Salle: Open Court.

Zalta, E. N. (Ed.) (2013). The Stanford encyclopedia of philosophy (Winter 2013 Edition). http://plato.stanford.edu/archives/win2013/.

Author Index

Subject Index

© Springer International Publishing Switzerland 2015
H. Barseghyan, *The Laws of Scientific Change*, DOI 10.1007/978-3-319-17596-6

Printed in the United States
By Bookmasters